Lecture Notes in Computer S

T0250655

Commenced Publication in 1973
Founding and Former Series Editors:
Gerhard Goos, Juris Hartmanis, and Jan van Leeuwen

Thomas Erlebach (Ed.)

Combinatorial and Algorithmic Aspects of Networking

Third Workshop, CAAN 2006
Chester, UK, July 2, 2006
Revised Papers

 Springer

Volume Editor

Thomas Erlebach
Department of Computer Science
University of Leicester
LE1 7RH, U.K.
E-mail: t.erlebach@mcs.le.ac.uk

Library of Congress Control Number: 2006937396

CR Subject Classification (1998): F.1.1, F.2.1-2, C.2, G.2.1-2, E.1

LNCS Sublibrary: SL 5 – Computer Communication Networks and
Telecommunications

ISSN   0302-9743
ISBN-10   3-540-48822-7 Springer Berlin Heidelberg New York
ISBN-13   978-3-540-48822-4 Springer Berlin Heidelberg New York

Springer is a part of Springer Science+Business Media

springer.com

© Springer-Verlag Berlin Heidelberg 2006

Typesetting: Camera-ready by author, data conversion by Scientific Publishing Services, Chennai, India
Printed on acid-free paper   SPIN: 11922377   06/3142   5 4 3 2 1 0

Preface

The Internet, because of its size, decentralized nature, and loosely controlled architecture, provides a hotbed of challenges that are amenable to mathematical analysis and algorithmic techniques. The primary goal of the 3rd Workshop on Combinatorial and Algorithmic Aspects of Networking (CAAN 2006) was to bring together mathematicians, theoretical computer scientists and network specialists in this fast-growing area that is an intriguing intersection of computer science, graph theory, game theory, and networks. CAAN 2006 took place on July 2, 2006 in Chester, UK, co-located with the 13th Colloquium on Structural Information and Communication Complexity (SIROCCO 2006). The two previous CAAN workshops were held during August 6-7, 2004 at the Banff International Research Station, Alberta, Canada and on August 14, 2005 in Waterloo, Ontario, Canada.

In response to the call for papers we received 22 submissions. Each submission was reviewed by four referees. Based on the reviews, the Program Committee selected ten papers for presentation at the workshop. The workshop program also featured an invited talk by David Peleg. This volume contains the contributed papers and an abstract of the invited talk.

We would like to thank the Organzing Committee of SIROCCO 2006, in particular Christoph Ambühl, Catherine Atherton, Leszek Gąsieniec and Prudence Wong, for all the organizational help that made it easy for us to arrange CAAN together with SIROCCO. Furthermore, we are grateful to Andrei Voronkov for providing the EasyChair conference system, which we used to manage the electronic submissions, the review process, and the electronic program committee meeting. It simplified our task significantly. Finally, we thank the invited speaker, the authors of the contributed papers, and all participants of CAAN 2006 for helping to make the workshop a success.

August 2006

Thomas Erlebach
Program Chair
CAAN 2006

Organization

Steering Committee

Andrei Broder	Yahoo! Inc., USA
Angèle Hamel	Wilfrid Laurier University, Canada
Srinivasan Keshav	University of Waterloo, Canada
Alejandro López-Ortiz	University of Waterloo, Canada
Rajeev Motwani	Stanford University, USA
Ian Munro	University of Waterloo, Canada

Program Committee

Christoph Ambühl	University of Liverpool
Holger Bast	MPI Saarbrücken
Gruia Calinescu	Illinois Institute of Technology
Andrea Clementi	University of Rome "Tor Vegata"
Colin Cooper	King's College London
Xiaotie Deng	City University of Hong Kong
Thomas Erlebach (Chair)	University of Leicester
Angèle Hamel	Wilfrid Laurier University
Samir Khuller	University of Maryland
Stavros Kolliopoulos	University of Athens
Danny Krizanc	Wesleyan University
Stefano Leonardi	University of Rome "La Sapienza"
Alejandro López-Ortiz	University of Waterloo
Christian Scheideler	TU München
Christian Schindelhauer	University of Freiburg
Angelika Steger	ETH Zürich
Csaba Tóth	Massachusetts Institute of Technology
Eli Upfal	Brown University

Additional Referees

Luca Becchetti	Yoo-Ah Kim	Maurizio Pizzonia
Rene Beier	Elias Koutsoupias	Guido Proietti
Tian-Ming Bu	Debapriyo Majumdar	Qi Qi
Hung Chim	Vangelis Markakis	Anuj Rawat
Miriam Di Ianni	Russell Martin	Gianluca Rossi
Angelo Fanelli	Julian Mestre	Wei Sun
Alexei Fishkin	Alfredo Navarra	Ingmar Weber
Leszek Gąsieniec	Francesco Pasquale	Prudence Wong
Min Jiang	Andrzej Pelc	Anders Yeo

Table of Contents

Invited Lecture

Contributed Papers

Recent Advances on Approximation Algorithms for Minimum Energy Range Assignment Problems in Ad-Hoc Wireless Networks

David Peleg[*]

Department of Computer Science and Applied Mathematics,
The Weizmann Institute of Science, Rehovot 76100, Israel
david.peleg@weizmann.ac.il

Ad-hoc wireless networks have no wired infrastructure. Instead, they consist of a collection of radio stations $S = \{1, 2, \ldots, n\}$ deployed in a given region and connected by wireless links. Each station is assigned a transmission range, and a station t can correctly receive the transmission of another station s if and only if t is within the range of s. The overall range assignment, $r : S \to R^+$, determines a (directed) transmission graph G_r. The transmission range of a station depends on the energy invested by the station. In particular, the power P_s required by a station s to correctly transmit data to another station t must satisfy the inequality $P_s \geq \text{dist}(s, t)^\alpha$, where $\text{dist}(s, t)$ is the Euclidean distance between s and t and $\alpha \geq 1$ is the *distance-power gradient*. The value of α may vary from 1 to more than 6 depending on the environment conditions at the location of the network (see [16]).

In order to allow an ad-hoc network to carry out certain basic communication paradigms, a fundamental design problem that needs to be solved is to calculate a transmission range assignment r such that (a) the corresponding transmission graph G_r satisfies a given connectivity property Π, and (b) the overall energy required to deploy the range assignment r is minimized. For any desired graph property Π, the resulting problem is denoted MIN-RANGE(Π).

We focus on two basic types of communication paradigms.

- *Broadcast* is a task initiated by a source station which has to disseminate a message to all stations in the wireless network. This task constitutes one of the main activities in real life multi-hop wireless networks [10].
- *Routing* is a task initiated by a source station which transmits a message intended to one particular destination station in the network.

To facilitate these two paradigms, the underlying transmission graph G_r is required to satisfy one of the following two properties, respectively.

B: Given a set of stations and a specific source station s, G_r has to contain a directed spanning tree rooted at s.

SC: Given a set of stations, G_r has to be *strongly connected*, i.e., contain a directed path from every station to every other station.

---

[*] Supported in part by a grant from the Israel Ministry of Science and Technology.

T. Erlebach (Ed.): CAAN 2006, LNCS 4235, pp. 1–4, 2006.

This characterization of the properties does not restrict the number of hops the communication might require. For quality of service purposes, it may be desirable to impose a bound h on the maximum number of hops in any communication path. This yields the following two variants.

$B[h]$: Given a set of stations and a specific source station s, G_r has to contain a directed spanning tree of depth at most h rooted at s.

$SC[h]$: Given a set of stations, G_r has to contain a directed path of at most h hops from every station to every other station.

We now review some known results on these problems. For broadcast, observe that if $\alpha = 1$, then the MIN-RANGE(B) problem is solvable in polynomial time. Moreover, in the 1-dimensional case (i.e., when the stations are placed on a line), the problem is solvable in polynomial time for any $\alpha \geq 1$ [6]. For $d \geq 2$ and $\alpha > 1$, however, MIN-RANGE(B) is NP-hard [5]. In [2,5] it is shown that whenever $\alpha \geq d$, the algorithm proposed in [10] based on constructing a minimum weight spanning tree achieves constant approximation. It is not known whether the problem admits a polynomial time approximation scheme.

Efficient solutions were given for MIN-RANGE($B[h]$) when h is constant. In particular, a polynomial-time algorithm for MIN-RANGE($B[h]$) for $h = 2$, based on a nontrivial dynamic program, is given in [1]. Moreover, the problem is given a polynomial-time approximation scheme for any fixed constant $h > 1$. For $\epsilon > 0$, the scheme has time complexity $O(n^\mu)$ where $\mu = O((\alpha 2^\alpha h^\alpha / \epsilon)^{\alpha^h})$.

For arbitrary h, an $O(hn^4)$-time exact algorithm for MIN-RANGE($B[h]$) on trees is presented in [18]. In addition they present a probabilistic $O(\log n \log \log n)$ approximation algorithm for MIN-RANGE($B[h]$) on a general metric space. These results are improved to a ratio of $O(\log n)$ in [3,12] independently. The existence of a polynomial time approximation scheme (or even a polynomial time constant ratio approximation algorithm) for arbitrary h is not known.

Turning to the strong connectivity property SC, we first remark that in the one-dimensional case, i.e., when the stations are located on the real line, the problem is polynomial. An $O(n^4)$-time algorithm for this problem is described in [13]. When the stations are spread in d-dimensional space ($d > 1$), finding an optimal solution for MIN-RANGE(SC) is NP-hard [9,13], and moreover, it is APX-hard for $d \geq 3$ [13]. On the positive side, the problem has a 2-approximation algorithm based on constructing a minimum spanning tree [13].

Finally, consider the bounded-hop strong connectivity requirement $SC[h]$. It is known that MIN-RANGE($SC[h]$) is NP-hard on general metric spaces for constant h [12]. For the 1-dimensional case where the stations of S are spread on the line, an $O(hn^3)$-time 2-approximation algorithm for $\alpha = 2$ and any $h > 0$ is described in [7]. In higher dimensions, lower and upper bounds are shown in [8] on the optimal cost for any 2-dimensional instances with distance power gradient $\alpha \geq 1$, where h is an arbitrary constant. It is also shown therein that when S is a family of *well-spread* instances (namely, the locations in S are suitably distributed), the MIN-RANGE($SC[h]$) problem on S admits a polynomial time approximation algorithm with constant ratio, i.e., MIN-RANGE($SC[h]$) is in APX. Additionally, it is shown that the MIN-RANGE($SC[h]$) problem with a uniform instance probability is in the class Av-APX.

For arbitrary h, a polynomial time approximation algorithm of ratio $O(n^2)$ for MIN-RANGE($SC[h]$) on trees and a randomized polynomial time approximation algorithm of ratio $O(n^2 \log n \log \log n)$ for the problem on general metric spaces were presented in [18]. This was improved in [3] to an approximation algorithm of ratio $O(\log n)$.

Finally, a polynomial time constant factor approximation algorithm for MIN-RANGE($SC[h]$) on general metrics is given in [12]. The approximation ratio of the algorithm is $\left(1/\left(\sqrt[h]{2}-1\right)\right)^{\alpha}(1+3^{\alpha})\left(3^{\alpha+1}\right)^{h-2}$. This is done by first considering a new variant of the classical *uncapacitated facility location* problem UFL [4,15,17], named the *hierarchical facility location* problem on metric powers, $\mathrm{HFL}_{\alpha}[h]$. This problem involves a set F of locations that may open a facility, subsets $D_1, D_2, \ldots, D_{h-1}$ of locations that may open an intermediate transmission station and a set D_h of locations of clients. Each client in D_h must be serviced by an open transmission station in D_{h-1} and every open transmission station in D_l must be serviced by an open transmission station on the next lower level, D_{l-1}. An open transmission station on the first level, D_1 must be serviced by an open facility. A cost c_{ij} is associated with assigning a station j on level $l \geq 1$ to a station i on level $l-1$. Also, for $i \in F$, a cost f_i is associated with opening a facility at location i. It is required to find a feasible assignment that minimizes the total cost.

A polynomial time constant ratio approximation algorithm is then established for the $\mathrm{HFL}_{\alpha}[h]$ problem, by solving a linear relaxation of the corresponding integer linear program and then using the filtering and rounding technique of [14,17]. Finally, the MIN-RANGE($SC[h]$) problem is reduced to $\mathrm{HFL}_{\alpha}[h]$, so that the constant approximation algorithm for $\mathrm{HFL}_{\alpha}[h]$ yields also an approximate solution for MIN-RANGE($SC[h]$).

References

1. C. Ambühl, A.E.F. Clementi, M. Di Ianni, N. Lev-Tov, A. Monti, D. Peleg, G. Rossi and R. Silvestri. Efficient Algorithms for Low-Energy Bounded-Hop Broadcast in Ad-Hoc Wireless Networks. In *Proc. 21st Symp. on Theoretical aspects of Computer Science*, pages 418–427, 2004.
2. G. Călinescu, X.Y. Li, O. Frieder, and P.J. Wan. Minimum-Energy Broadcast Routing in Static Ad Hoc Wireless Networks. In *Proc. 20th INFOCOM*, pages 1162–1171, 2001.
3. J. Chlebikova, D. Ye and H. Zhang. Assign Ranges in General Ad-Hoc Networks. In *Proc. 1st Conf. on Algorithmic Applications in Management*, Xian, China, pages 411–421, 2005.
4. F.A. Chudak. Improved approximation algorithm for uncapacitated facility location problem. In *Proc. 6th Conf. on Integer Programming and Combinatorial Optimization*, pages 180–194, 1998.
5. A.E.F. Clementi, P. Crescenzi, P. Penna, G. Rossi, and P. Vocca. On the Complexity of Computing Minimum Energy Consumption Broadcast Subgraphs. In *Proc. 18th Symp. on Theoretical Aspects of Computer Science*, pages 121–131, 2001.

6. A.E.F. Clementi, M. Di Ianni, and R. Silvestri. The Minimum Broadcast Range Assignment Problem on Linear Multi-Hop Wireless Networks. *Theoretical Computer Science* 299, (2003), 751–761.

7. A. Clementi, A. Ferreira, P. Penna, S. Perennes, and R. Silvestri. The minimum range assignment problem on linear radio networks. In *Proc. 8th European Symp. on Algorithms*, pages 143–154. 2000.

8. A. Clementi, A. Ferreira, P. Penna, S. Perennes, and R. Silvestri. The power range assignment problem in radio networks on the plane. In *Proc. 17th Symp. on Theoretical Aspects of Computer Science*, pages 651–660, 2000.

9. A.E.F. Clementi, P. Penna, and R. Silvestri. Hardness results for the power range assignment problem in packet radio networks. In *Proc. 2nd Workshop on Approximation Algorithms for Combinatorial Optimization Problems*, pages 197–208, 1999.

10. A. Ephremides, G.D. Nguyen, and J.E. Wieselthier. On the Construction of Energy-Efficient Broadcast and Multicast Trees in Wireless Networks. In *Proc. 19th INFOCOM*, pages 585–594, 2000.

11. J. Fakcharoenphol, S. Rao, and K. Talwar. A tight bound on approximating arbitrary metrics by tree metrics. In *Proc. 35th ACM Symp. on Theory of Computing*, pages 448–455, 2003.

12. E. Kantor and D. Peleg. Approximate Hierarchical Facility Location and Applications to the Shallow Steiner Tree and Range Assignment Problems. *Proc. 6th Conf. on Algorithms and Complexity*, pages 211–222, 2006.

13. L. M. Kirousis, E. Kranakis, D. Krizanc, and A. Pelc. Power Consumption in Packet Radio Networks. *Theoretical Computer Science* 243, (2000), 289–305.

14. J.H. Lin and J.S. Vitter. ε–approximations with small packing constraint violation. In *Proc. 24th ACM Symp. on Theory of Computing*, pages 771–782, 1992.

15. M. Mahdian, Y. Ye, and J. Zhang. A 1.52-approximation algorithm for the uncapacitated facility location problem. In *Proc. 5th Workshop on Approximation Algorithms for Combinatorial Optimization Problems*, pages 229–242, 2002.

16. K. Pahlavan and A. Levesque. *Wireless information networks*. Wiley-Interscience, 1995.

17. B.D. Shmoys, E. Tardos, and Aardal K. Approximation algorithms for facility location problems. In *Proc. 29th ACM Symp. on Theory of Computing*, pages 265–274, 1997.

18. D. Ye and H. Zhang. The range assignment problem in static ad-hoc networks on metric spaces. In *Proc. 11th Colloq. on Structural Information and Communication Complexity*, pages 291–302, 2004.

The Price of Anarchy in Selfish Multicast Routing

(Extended Abstract)

Andreas Baltz⋆, Sandro Esquivel, Lasse Kliemann⋆, and Anand Srivastav

Institut für Informatik, CAU Kiel
Christian-Albrechts-Platz 4
24118 Kiel
{aba, sae, lki, asr}@numerik.uni-kiel.de

Abstract. We study the price of anarchy for selfish multicast routing games in directed multigraphs with latency functions on the edges, extending the known theory for the unicast situation, and exhibiting new phenomena not present in the unicast model. In the multicast model we have N commodities (or player classes), where for each $i = 1, \ldots, N$, a flow from a source s_i to a finite number of terminals $t_i^1, \ldots, t_i^{k_i}$ has to be routed such that every terminal t_i^j receives flow $n_i \in \mathbb{R}_{\geq 0}$.

One of the significant results of this paper are upper and lower bounds on the price of anarchy for edge latencies being polynomials of degree at most p with non-negative coefficients. We show an upper bound of $(p+1) \cdot \frac{\nu^{p+1}}{\nu^*}$ in some variants of multicast routing. We also prove a lower bound of ν^p, so we have upper and lower bounds that are tight up to a factor of $(p+1)\nu$. Here, ν and ν^* are network and strategy dependent parameters reflecting the maximum/minimum consumption of the network. Both are 1 in the unicast case. Our lower bound of ν^p, where in the general situation we have $\nu > 1$, shows an exponential increase compared to the Roughgarden bound of $O(p/\ln p)$ for the unicast model. This exhibits the contrast to the unicast case, where we have Roughgarden's (2002) result that the price of anarchy is independent of the network topology. To our knowledge this paper is the first thorough study of the price of anarchy in the multicast scenario. The approach may lead to further research extending game-theoretic network analysis to models used in applications.

1 Introduction

Multicast routing in communication networks is a natural and practically relevant extension of the so far quite well studied unicast routing. Among the applications of multicast routing are the transmission of music, movies, conferences, or any other popular content, that is requested by several customers at a time. A formal description of our multicast routing model needs many technical definitions. We keep the introduction on a more informal level and refer the reader to Sect. 2 for all necessary details.

⋆ Supported by Deutsche Forschungsgemeinschaft.

T. Erlebach (Ed.): CAAN 2006, LNCS 4235, pp. 5–18, 2006.

Problem Formulation. An instance of selfish multicast routing consists of a directed multigraph $G = (V, E)$, where the edges are also called links, a set of N player classes, called commodities, where commodity i is characterized by a source s_i and terminals (or sinks) $t_i^1, \ldots, t_i^{k_i}$, and a (flow) demand of $n_i \in \mathbb{R}_{\geq 0}$. The links are each equipped with a latency function $l_e : \mathbb{R}_{\geq 0} \longrightarrow \mathbb{R}_{\geq 0}$. For commodity i, a set $S = \{P_1, \ldots, P_{k_i}\}$ where P_j is an s_i-t_i^j-path, is called a strategy. The task is to realize for every commodity i, a flow in the network from s_i to all terminals $t_i^1, \ldots, t_i^{k_i}$, satisfying the demand n_i for every terminal. We think of the demand as being under control of infinitely many players, each controlling a negligible fraction and selfishly trying to find the fastest route for it. This game-theoretic model is known as the Wardrop model.

In the unicast model, $k_i = 1$ for all i, so we have a collection of single source/single sink commodities, and every strategy S consists of one path only. In the multicast case there are two different ways to route the flow $f(S)$ assigned to a strategy S for commodity i: either, we route $f(S)$ on each path, which is the usual notion of flows satisfying the Kirchhoff conservation law (here shortly called *conservation flow*), or we allow multiple duplication of flow at certain nodes: a link which serves several, say r, terminals in a strategy, i.e., a link contained in r paths of that strategy, only needs to transmit the data once, not r times. That is because the data can later be duplicated to serve all terminals. In this way, the congestion on the links can be reduced. We call such a flow *duplication flow*.

The cost of a flow is defined by $\mathsf{SC}(f) = \sum_{S \in \mathfrak{S}} l_S(f) f(S)$, where $\mathfrak{S}$ is the set of all strategies of all commodities, $f(S) \leq n_i$ is the portion of the demand that by the decision of the selfish players has been allocated to strategy S, and $l_S(f)$ is the strategy latency for S. We study four different definitions for l_S, which all coincide in the unicast case. Together with the two types of flows (conservation and duplication), we thus have 8 variants of multicast. The price of anarchy for a multicast instance $\mathcal{I}$ is $\rho(\mathcal{I}) = \sup_f \frac{\mathsf{SC}(f)}{\mathsf{SC}(f^*)}$, where f ranges over all Nash equilibria and f^* is an optimal flow. A Nash equilibrium is a flow in which no player (meaning: no portion of the flow, however small) has an incentive to unilaterally deviate from his current strategy.

Previous and Related Work. By the pioneering work of Roughgarden [1] and Roughgarden and Tardos [2] we know that $\rho(\mathcal{I})$ in the unicast model for latency functions being polynomials of degree p, is bounded from above by $O(\frac{p}{\ln p})$ (and is $\frac{4}{3}$ for $p = 1$). As already an example of a 2-parallel links network has a price of anarchy of $\frac{4}{3}$ (for $p = 1$), the surprising conclusion is that it is *independent* of the network topology from a worst-case point of view [1, Sec. 3.4].

Our Results. A solid foundation for the analysis of multicast routing games is given in Sect. 2. In Sect. 2.1 we introduce a concise model, and in Sect. 2.2 we show, using results on variational inequalities, the existence of Nash equilibria.

As a main result, we show in Sect. 3 that the price of anarchy in multicast routing may depend heavily on the network topology and the strategies. Certain edges of the graph may be utilized under certain strategies more than others, although the players on those strategies are not charged for this. On the other hand, some strategies may depend highly on some edges but only contribute a small amount to their utilization. See Remark 2 for a more detailed discussion of this. To capture the effects of this phenomenon, which does not occur in unicast routing, we introduce for each edge and strategy an integer called the *consumption*. Moreover we introduce two new invariants for a graph G and a set of strategies $\mathfrak{S}$, which we call *maximum (resp. minimum) consumption number*, $\nu = \nu(G, \mathfrak{S})$ resp. $\nu^* = \nu^*(G, \mathfrak{S})$. We have $\nu = 1 = \nu^*$ in unicast routing. We show in Sect. 3.3 that in two variants of multicast for polynomial latency functions of degree p (where we will always assume non-negative coefficients), the price of anarchy is at most $(p + 1)\frac{\nu^{p+1}}{\nu^*}$ and in Sect. 3.2 provide a lower bound for one of these variants of ν^p (with $\nu^* = 1$). So, we have here a gap of $(p + 1)\nu$.

We then present (also in Sect. 3.2) a multicast instance with price of anarchy at least ν^p. As in general $\nu > 1$, the ν^p bound is exponentially larger than the corresponding unicast bound of $O(\frac{p}{\ln p})$. This is surprising (and disappointing from the point of view of a company running the network). For instances using the advantages of duplication flows in order to de-load high-latency links, the cost of the global optimum decreases drastically, but unfortunately, due to selfish behavior, the users grab (greedily) certain links without a look ahead and block them out, so that the cost of the Nash equilibrium still stands high. For other definitions of strategy latency, in Sect. 4 we are able to prove that results from non-atomic congestion games, i.e., bounds of the form $O(\frac{p}{\ln p})$, carry over.

Open Problems. A couple of interesting open problems arise from this paper. For example, can the $(p + 1)\nu$ factor gap between upper and lower bound be closed? Can exponentially high prices of anarchy be reduced by taxation schemes? It would also be interesting to consider polynomial time algorithms for the computation of equilibria.

An ambitious task would be to study multicast for information flows with duplication *and* coding facilities of the network. Such networks are the state-of-the-art in today's engineering designs. Our work can be considered as a first step in this direction.

2   Basics of Multicast Routing

2.1   Model and Instances

An instance of selfish multicast routing consists of the following.

- A directed multigraph $G = (V, E)$. The edges are also called *links*.

- A set of N *player classes* (or *user classes*). Sometimes, player classes are also called *commodities*. Each player class is characterized by a demand n_i and a vector of vertices $(s_i; t_i^1, \ldots, t_i^{k_i})$, where s_i is the source, and the $t_i^1, \ldots, t_i^{k_i}$ are the terminals.

• The demand n_i is supposed to be routed from s_i to each of the terminals $t_i^1, \ldots, t_i^{k_i}$. We think of the demand as being under control of infinitely many players, each of them controlling a negligible amount of it. This is the well-known Wardrop model (see, e.g., [1, Sec. 2.2]), which will become clearer when we define flows and Nash equilibria below.

• Each link $e \in E$ in the graph is equipped with a latency function $l_e : \mathbb{R}_{\geq 0} \longrightarrow \mathbb{R}_{\geq 0}$. We always assume each latency function to be non-decreasing and *standard* [1]. This means that it is differentiable and $\xi \mapsto l_e(\xi)\xi$ is convex.

If an amount ξ of traffic is to be routed through the link e, each unit of flow will take $l_e(\xi)$ time to traverse e. Hence we have a total latency of $l_e(\xi)\xi$ on that link.

• For $i \in [N] = \{1, \ldots, N\}$, we call a set of paths $S := \{P_1, \ldots, P_{k_i}\}$ where P_j is a path connecting s_i with t_i^j for $j \in [k_i]$, a *strategy*. Note that for unicast routing $k_i = 1$ for all i. The set of all strategies we wish to allow for player class i is denoted by $\mathfrak{S}_i$. We assume[1] that $\mathfrak{S}_i \cap \mathfrak{S}_j = \emptyset$ for all $i, j \in [N]$. Let $\mathfrak{S} := \bigcup_{i \in [N]} \mathfrak{S}_i$.

• An *action distribution* (according to [2]), simply called *flow*, is a map $f : \mathfrak{S} \longrightarrow \mathbb{R}_{\geq 0}$ such that all the demands are met, i.e., $\sum_{S \in \mathfrak{S}_i} f(S) = n_i \quad \forall i \in [N]$. A flow can be understood as a partition of each of the real intervals $[0, n_i]$. Each of these intervals represents the continuum of infinitely many players of the corresponding player class. The quantity $f(S)$ gives, for each $S \in \mathfrak{S}_i$, the portion of demand that by the decision of the players from that class is routed according to that particular strategy S.

As described in the introduction, the routing of a flow in the multicast model can be done in two different ways: we can route the demand with flows in the usual sense (conservation flows) or with flows allowing duplication (duplication flows).

• Let $e \in E$ and $S \in \mathfrak{S}$. We define the *consumption* of e under S as $c(e, S) := |\{P \in S; e \in P\}|$, i.e., the consumption is the number of paths in S traversing e, or in other words, the number of terminals served via e in this strategy.

• The *congestion* f_e of a link e with respect to a flow f is the amount of traffic that link e has to process. The total latency of a link e hence is $l_e(f_e)f_e$. Each instance defines the congestion in one of the following ways, depending on whether we have conservation flows or duplication flows.

$$f_e := \begin{cases} \sum_{S \in \mathfrak{S}(e)} c(e, S) f(S) & \text{conservation flows} \\ \sum_{S \in \mathfrak{S}(e)} f(S) & \text{duplication flows} \end{cases} \tag{1}$$

Here, $\mathfrak{S}(e)$ denotes the set of all strategies that contain a path which in turn contains e.

• We denote by $l_S(f)$ the so-called latency of strategy S with respect to a flow f. In unicast $l_S(f)$ is simply the sum of the latencies in the single path of which

---

[1] Otherwise we have to treat $\mathfrak{S}$ as a multiset.

the strategy S consists[2]. Let $S = \{P_1, \ldots, P_{k_i}\} \in \mathfrak{S}$. As in [1], the latency of a path P under f is defined by

$$l_P(f) := \sum_{e \in P} l_e(f_e).$$

By $E(S)$ we denote the union of the edges in all the paths in S. Note that we consider $E(S)$ *not* as a multiset, so edges do not appear multiple times even if they lie in several paths.

We introduce the following four definitions of the latency of a strategy S.

$$l_S^{\text{edges}}(f) := \sum_{e \in E(S)} l_e(f_e) \ , \quad l_S^{\text{paths}}(f) := \sum_{P \in S} l_P(f)$$

$$l_S^{\text{paths avg}}(f) := \frac{1}{|S|} l_S^{\text{paths}}(f) \ , \quad l_S^{\max}(f) := \max_{P \in S} l_P(f) \tag{2}$$

An instance of selfish multicast routing includes one of these strategy latency functions.

The latency of a strategy $l_S(f)$ is the latency that all players experience who choose strategy S. It can hence be thought of as a kind of equivalent to what is known as utility or payoff function in other game-theoretic settings.

Remark 1. 1. For unicast routing, all four definitions in (2) coincide.
2. It is easy to see that $l_S^{\text{paths}}(f) = \sum_{e \in E(S)} c(e, S) l_e(f_e)$.

2.2 Nash Equilibria, Social Cost, Price of Anarchy

A flow f is called a *Nash equilibrium* (sometimes abbreviated *NE*), if

$$f(S_1) > 0 \implies l_{S_1}(f) \le l_{S_2}(f) \quad \forall S_1, S_2 \in \mathfrak{S}_i \quad \forall i \in [N] \ . \tag{3}$$

Hence, in a Nash equilibrium, only minimum-latency strategies are used, since then no player has an incentive to choose a different strategy (provided the rest of the players keep their current decision). If each l_S is continuous (which will be the case during all our studies), then the game admits at least one Nash equilibrium. This follows from the characterization of Nash equilibria as the solutions to a certain variational inequality (see Thm. 1 and the discussion after that).

We define the *social cost* of a flow f as

$$\text{SC}(f) := \sum_{S \in \mathfrak{S}} l_S(f) f(S) \ .$$

The social cost captures the overall performance of the system for a given flow f. We will always assume that our instances admit a flow f^* with minimum social cost and that $\text{SC}(f^*) > 0$. Existence is guaranteed if all l_S are continuous

---

[2] The maximum over all links in the path has also been studied [3].

(which will be the case in our studies), because the set of flows is compact. For an instance $\mathcal{I}$ of selfish multicast routing with optimal flow f^*, define the price of anarchy by

$$\rho(\mathcal{I}) := \sup_{f \text{ is NE}} \frac{\mathsf{SC}(f)}{\mathsf{SC}(f^*)} \ .$$

Nash equilibria have a very simple structure, as seen in the following proposition. The proof for this is straightforward.

Proposition 1. *Let f be a Nash equilibrium. Then, for every $i \in [N]$, there exists a real number $l_i(f)$ such that $l_S(f) = l_i(f)$ for all $S \in \mathfrak{S}_i$, whenever $f(S) > 0$, and no strategy in $\mathfrak{S}_i$ has latency less than $l_i(f)$.*

Corollary 1. *Let f be a Nash equilibrium. Then $\mathsf{SC}(f) = \sum_{i \in [N]} l_i(f) n_i$.*

We now aim for further characterizations of Nash equilibria. Let $f, \tilde{f}$ be flows. Define $\mathsf{SC}^f(\tilde{f}) := \sum_{S \in \mathfrak{S}} l_S(f)\tilde{f}(S)$. The first part of the following theorem is well-known for the unicast case, see, e.g., [1, Lem. 3.3.7] or [4] and the references therein. The whole theorem also holds in a more general context than multicast routing, for it (and its proof) does not require the notion of congestion.

Theorem 1.   *1. Let f be a flow. Then f is a Nash equilibrium if and only if we have*

$$\mathsf{SC}^f(\tilde{f}) \geq \mathsf{SC}(f) \quad \text{for all flows } \tilde{f} \ . \tag{4}$$

2. Let each l_S be continuous. Then the multicast game admits at least one Nash equilibrium.

Proof. We refer the reader to the full version of this paper for the proof of 1). For 2) note that (4) is equivalent to $\sum_{S \in \mathfrak{S}} l_S(f)(\tilde{f}(S) - f(S)) \geq 0$ for all flows $\tilde{f}$. This is a well-studied variational inequality. It has been shown in [5] with deep results from the index theory of vector fields that it admits at least one solution. □

Note that all strategy latencies from (2) are continuous, because we only consider standard latency functions and because the congestion is a continuous mapping. Hence, all our multicast games admit at least one Nash equilibrium.

In the rest of the paper we investigate the price of anarchy for conservation flows resp. duplication flows and the four strategy latencies from (2). These are 8 cases. In Sect. 3 we show for 5 of them that the price of anarchy depends on the network topology, while in two other cases it does not (Sect. 4).

3   Price of Anarchy Dependent on the Network Topology

For a directed graph G and a set of strategies $\mathfrak{S}$ we define the *minimum and maximum consumption number* as

$$\nu^*(G, \mathfrak{S}) := \min_{S \in \mathfrak{S}} \min_{e \in E(S)} c(e, S) \ , \quad \nu(G, \mathfrak{S}) := \max_{S \in \mathfrak{S}} \max_{e \in E(S)} c(e, S) \ .$$

We write just ν^* and ν, if we are dealing with only one instance at a time. In a unicast situation there is only one path in each strategy and hence $\nu^*(G, \mathfrak{S}) = 1 = \nu(G, \mathfrak{S})$.

3.1   Latency $l^{\text{paths avg}}$ and l^{max}

Consider the instance $\mathcal{I}^{r,R,N}$ in Fig. 1. In this instance, we have $N + 1$ player classes, all sharing the source s.

- For each $i \in [N]$, player class i has one terminal t_i. The demand is r, for a fixed $r \in \mathbb{R}_{\geq 0}$.
- Player class $N + 1$ has N terminals $t_1, \dots t_N$. This class has demand R, $R \in \mathbb{R}_{\geq 0}$ fixed.

There are two links between s and each t_i, one with constant latency $r + R$ and one with latency function $l : \xi \mapsto \xi$. Defining a strategy in this example for player class $N + 1$ means picking for each t_i either the ξ-link or the link with constant latency. Conservation flows and duplication flows coincide, since each strategy consists of edge disjoint paths.

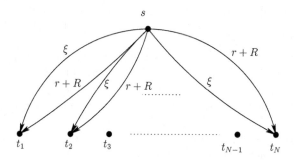

Fig. 1. Example $\mathcal{I}^{r,R,N}$. The values r and R are non-negative real numbers. ξ denotes the latency function $l : \xi \mapsto \xi$.

Theorem 2. *By choosing r sufficiently small and R and N sufficiently large, the price of anarchy in the example in Fig. 1 can be made arbitrarily high (although we still have linear latency functions). More precisely, we have for large N*

$$\rho(\mathcal{I}^{r,R,N}) \geq (1 + \frac{R}{r})(1 - o(1)) \ .$$

This holds for all four combinations of conservation flows and duplication flows on the one hand, and $l^{\text{paths avg}}$ and l^{max} on the other hand.

Proof. Let f be the flow where all players from all classes choose the links with latency function $\xi \mapsto \xi$. We then have $l_S^{\text{paths avg}}(f) = r + R$ for all strategies S. The same holds for l^{max}. Hence, f is a Nash flow for $l^{\text{paths avg}}$ and l^{max}. By

Cor. 1 we have $\mathsf{SC}(f) = \left(\sum_{i \in [N]}(r+R)r\right) + (r+R)R = Nr(r+R) + (r+R)R$.

For comparison, take $\widetilde{f}$ as the flow in which all players from classes 1 to N stick to the ξ-links, but all players from class $N+1$ take the links with constant latency $r+R$. We then have (for $l^{\text{paths avg}}$ and l^{max}) $\mathsf{SC}(\widetilde{f}) = \left(\sum_{i \in [N]} r \cdot r\right) + (r+R)R = Nr^2 + (r+R)R$. Hence, the price of anarchy is at least

$$\rho(\mathcal{I}^{r,R,N}) \geq \frac{Nr(r+R) + (r+R)R}{Nr^2 + (r+R)R} \quad . \tag{5}$$

Since the term from (5) tends to $1 + \frac{R}{r}$ as $N \to \infty$, the claim follows. □

3.2 Duplication Flows and l^{paths}

Theorem 3. *There are examples of selfish multicast routing instances with duplication flows using l^{paths} and latency functions that are polynomials of degree $\leq p$ (with non-negative coefficients) where the price of anarchy is at least ν^p.*

Proof. Consider the instance in Fig. 2. We are given one player class with k terminals $t^1, \ldots, t^k$. The demand is 1. Note that in this example, $\nu = k$. The latency functions on the links on the upper path are identically 0, and these links may be used in any direction. The links of the form (s, t^j) all have latency function $\xi \mapsto \xi^p$ for a fixed p.

A Nash equilibrium f is achieved if all players use the tree consisting of the edges $(s, t^1), \ldots, (s, t^k)$. This flow has social cost $\mathsf{SC}(f) = k$.

Now, a better flow $\widetilde{f}$ is given as follows. For every $j \in [k]$ let a fraction of $\frac{1}{k}$ users 'inject' their flow into the upper path via the edge (s, t^j). Each of these strategies has latency $k\frac{1}{k^p} = k^{1-p}$, because the edge (s, t^j) has congestion $\frac{1}{k}$ and hence a latency of $\frac{1}{k^p}$, and this edge is contained in k paths. Because there are k such strategies, each of them carrying $\frac{1}{k}$ units of flow, we have a social cost of $\mathsf{SC}(\widetilde{f}) = kk^{1-p}\frac{1}{k} = k^{1-p}$. Hence the price of anarchy ρ of this instance is at least $\rho \geq \frac{\mathsf{SC}(f)}{\mathsf{SC}(\widetilde{f})} = kk^{p-1} = k^p$. □

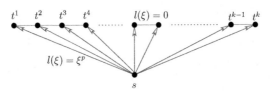

Fig. 2. The links in the upper path can be used in both directions and have latency 0

In Thm. 7 we will see that the same instance for conservation flows (and l^{paths}) has a price of anarchy bounded by $O(p)$. So, we have an exponential increase in the price of anarchy when we switch to duplication. This is due to the much better optimum which takes advantage of duplication. However, the ν^p bound from Thm. 3 says that due to uncoordinated behavior of the players, the benefit of duplication is completely neglected.

Summarizing, we considered the cases of conservation flows and $l^{\text{paths avg}}$ or $l^{\max}$, duplication flows and $l^{\text{paths avg}}$ or $l^{\max}$, and duplication flows and l^{paths}. For these cases, the price of anarchy — even from a worst-case point of view — is dependent on the network topology and the set of strategies, i.e., we can construct instances with arbitrarily high prices of anarchy for some fixed class of latency functions.

3.3   Upper Bounds for Polynomial Latency Functions

We consider the following two cases: a) conservation flows and l^{edges}, b) duplication flows and l^{paths}. We also restrict ourselves to polynomial latency functions with non-negative coefficients, i.e.,

$$l_e(\xi) = \sum_{j=0}^{p} \lambda_{e,j}\xi^j \quad \forall e \in E \quad \forall \xi \in \mathbb{R}_{\geq 0} \ , \tag{6}$$

where $\lambda_{e,j} \in \mathbb{R}_{\geq 0}$ for all e and j. The following holds for case a) as well as b).

Proposition 2. *Let $f, \widetilde{f}$ be flows. Then*

$$\sum_{e\in E} l_e(f_e)\widetilde{f}_e \leq \nu^p \sum_{e\in E} l_e(\widetilde{f}_e)\widetilde{f}_e + \frac{p}{(p+1)\nu} \sum_{e\in E} l_e(f_e)f_e \ .$$

Conservation Flows and l^{edges}. We first prove an upper bound on the total latency $\sum_{e\in E} l_e(f_e)f_e$, which holds for any kind of latency functions. The proofs will be given in a full version of this paper.

Lemma 1. *For every flow f we have $\sum_{e\in E} l_e(f_e)f_e \leq \nu \cdot \mathsf{SC}(f)$.*

The proof of the following theorem uses methods from [4, Th. 2.1.].

Theorem 4. *Let $\mathcal{I}$ be an instance of selfish multicast routing using conservation flows, l^{edges}, and polynomial latency functions of degree at most p and non-negative coefficients on the edges. Then $\rho(\mathcal{I}) \leq (p+1)\frac{\nu^{p+1}}{\nu^*}$.*

Proof. Let f be a Nash equilibrium and $\widetilde{f}$ some flow. By Thm. 1 we get

$$\mathsf{SC}(f) \leq \mathsf{SC}^f(\widetilde{f}) = \sum_{S\in\mathfrak{S}} l_S^{\text{edges}}(f)\widetilde{f}(S) = \sum_{S\in\mathfrak{S}} \sum_{e\in E(S)} l_e(f_e)\widetilde{f}(S)$$

$$= \sum_{e\in E} \sum_{S\in\mathfrak{S}(e)} l_e(f_e)\widetilde{f}(S) = \frac{1}{\nu^*} \sum_{e\in E} l_e(f_e) \sum_{S\in\mathfrak{S}(e)} \nu^*\widetilde{f}(S)$$

$$\leq \frac{1}{\nu^*} \sum_{e\in E} l_e(f_e) \sum_{S\in\mathfrak{S}(e)} c(e,S)\widetilde{f}(S) = \frac{1}{\nu^*} \sum_{e\in E} l_e(f_e)\widetilde{f}_e \ .$$

By Prop. 2 and Lem. 1 we get

$$\frac{1}{\nu^*}\sum_{e\in E}l_e(f_e)\widetilde{f}_e \leq \frac{1}{\nu^*}\nu^p\sum_{e\in E}l_e(\widetilde{f}_e)\widetilde{f}_e + \frac{1}{\nu^*}\frac{p}{(p+1)\nu}\sum_{e\in E}l_e(f_e)f_e$$

$$\leq \frac{1}{\nu^*}\nu^p\sum_{e\in}l_e(\widetilde{f}_e)\widetilde{f}_e + \frac{p}{(p+1)\nu}\sum_{e\in E}l_e(f_e)f_e \leq \frac{\nu^{p+1}}{\nu^*}\mathsf{SC}(\widetilde{f}) + \frac{p}{p+1}\mathsf{SC}(f) \ .$$

It follows that $\frac{1}{p+1}\mathsf{SC}(f) \leq \frac{\nu^{p+1}}{\nu^*}\mathsf{SC}(\widetilde{f})$, hence $\frac{\mathsf{SC}(f)}{\mathsf{SC}(\widetilde{f})} \leq (p+1)\frac{\nu^{p+1}}{\nu^*}$. The claim follows by using an optimal flow as $\widetilde{f}$. $\square$

For linear latency functions, i.e., $p = 1$, we can along similar lines show a slightly better bound of $\frac{4}{3}\cdot\frac{\nu^2}{\nu^*}$.

Duplication Flows and l^{paths}. In a similar way as in the previous section, 3.3, we can prove for duplication flows and l^{paths}:

Lemma 2. *For every flow f we have $\nu^*\sum_{e\in E}l_e(f_e)f_e \leq \mathsf{SC}(f)$.*

Theorem 5. *Let $\mathcal{I}$ be an instance of selfish multicast routing using duplication flows, l^{paths}, and polynomial latency functions of degree at most p and non-negative coefficients on the edges. Then $\rho(\mathcal{I}) \leq (p+1)\frac{\nu^{p+1}}{\nu^*}$.*

4   Price of Anarchy Independ. of the Network Topology

The proofs in this section are based on identifying the given multicast game with a *non-atomic congestion game with separable latencies (separable NCG for* short). These games are well studied [2]. We first describe known results on such games, and their connection to our multicast game.

4.1   Non-atomic Congestion Games with Separable Latencies

A separable NCG is defined similar to our multicast game, however, there is no graph required. We are simply given a set E of elements. Each of the elements has a latency function l_e. A strategy is a subset of E. By $\mathfrak{S}(e)$ we denote all strategies containing element e. Crucial for a NCG are its definitions of congestion and strategy latency. For each $e \in E$ and $S \in \mathfrak{S}$ let a_{eS} be some non-negative real number. For a given flow, the congestion of an edge e resp. latency of a strategy S is defined by

$$f_e := \sum_{S\in\mathfrak{S}(e)} a_{eS}f(S) \ , \ \text{resp.} \ l_S(f) := \sum_{e\in S}a_{eS}l_e(f_e) \ .$$

Nash equilibria, social cost and price of anarchy are defined as in our multicast games. An important point is that the factors a_{eS} appear both in congestion and in strategy latency. This allows us [2, Prop. 2.8] to write the social cost as $\mathsf{SC}(f) = \sum_{e\in E}l_e(f_e)f_e$.

Remark 2. Comparing this to multicast games with duplication flows and l^{paths}, we note that they do not have this kind of symmetry. Instead, they carry factors like the a_{eS} (the consumptions, in fact), but only in the definition of strategy latency. These factors do not appear in the congestion. As we have seen in Sect. 3.2, the lack of such a symmetry can lead to high prices of anarchy, that may even depend on the network topology. Other variants, like conservation flows and l^{edges} have the consumption as a factor in the congestion, but not in the strategy. Having high factors in the strategy and relatively small factors in the congestion means that players choosing that strategy suffer highly under a high congestions on this edge, but do not contribute much to that congestion. This can happen in multicast with duplication flows and l^{paths} with respect to edges that serve many terminals (in the chosen strategy).

On the other hand, high factors in the congestion but only relatively small factors in the strategy latency means that although the players on this strategy induce a high utilization on a link, they are not charged accordingly. This can happen in multicast with conservation flows and l^{edges} with respect to edges that serve many terminals (in the chosen strategy).

These observations give rise to the question about a new, more general class of congestion games, in which we have different factors in the strategy latencies and the congestion.

The price of anarchy for a separable NCG with standard latency functions is well understood [2]. Let $\mathcal{L}$ be a class of standard latency functions that contains at least one non-zero function. We call such a class also *standard*. Note, that we will always assume that our multicast games have at least one link with a non-zero latency function. Hence, if we consider the class of all latency functions for a given instance, this class is automatically standard. Now, it is easy to see that for each $l \in \mathcal{L}$ there exists $\beta_l : \mathbb{R}_{\geq 0} \longrightarrow \mathbb{R}_{\geq 0}$ such that $l^*(\beta_l(\xi)) = l(\xi)$ for all $\xi \in \mathbb{R}_{\geq 0}$, where $l^*(\xi) := l'(\xi)\xi + l(\xi)$. The *anarchy value* of $\mathcal{L}$ [2, Def. 4.3], see also [1, Def. 3.3.2], is

$$\alpha(\mathcal{L}) := \sup_{0 \neq l \in \mathcal{L}, \, \xi > 0, \, l(\xi) > 0} \frac{l(\xi)\xi}{l(\xi)\xi + l(\beta_l(\xi))\beta_l(\xi) - l(\xi)\beta_l(\xi)} .$$

The amazing property of separable NCGs with latency functions from a standard class $\mathcal{L}$ is that their price of anarchy can be upper-bounded in terms $\alpha(\mathcal{L})$ — a parameter that is independent of the structure of $\mathfrak{S}$.

Theorem 6 (Roughgarden, Tardos [2]). *Let Γ be a separable NCG with latency functions from a standard class $\mathcal{L}$. Then $\rho(\Gamma) \leq \alpha(\mathcal{L})$.*

For polynomial latency functions the anarchy values are well-known.

Corollary 2 ([2]). *Let Γ be a separable NCG with latency functions that are polynomials of degree at most p with non-negative coefficients. Then*

$$\rho(\Gamma) \leq \left(1 - p(p+1)^{-(p+1)/p}\right)^{-1} .$$

For linear latency functions, i.e., if $p = 1$, the quantity on the right hand side is $\frac{4}{3}$. For $p \to \infty$, it is $\Theta(\frac{p}{\ln p})$, yielding a rough bound of $O(p)$.

The bound in Thm. 6 is tight in the following sense: let $\mathcal{L}$ be a standard class containing the constant functions[3]. Then we can find for every $\epsilon > 0$, a separable NCG Γ with latency functions from $\mathcal{L}$ and $\rho(\Gamma) > \alpha(\mathcal{L}) - \epsilon$. This means that — from a worst-case point of view — the price of anarchy in separable NCGs is independent of the strategy structure and the demands.

4.2   Duplication Flows and l^{edges}, Conservation Flows and l^{paths}

Theorem 7. *Let $\mathcal{L}$ be standard. Let $\mathcal{I}$ be an instance of selfish multicast routing with latency functions from $\mathcal{L}$, then we have $\rho(\mathcal{I}) \leq \alpha(\mathcal{L})$ if we consider a) duplication flows and use l^{edges}, or b) conservation flows and use l^{paths}.*

Proof. We only consider case a), i.e., duplication flows and strategy latencies l^{edges}, and refer to the full version for the second case. The first case can be modeled as a separable NCG as follows: we take the links in E as elements and as latency functions for the elements we take the latency functions of the links. A strategy S is identified with $E(S)$, so that it can be seen as a subset of the set of elements. For all $e \in E$ and $S \in \mathfrak{S}$ set

$$a_{eS} := \begin{cases} 1 & \text{if } e \in E(S) \\ 0 & \text{otherwise.} \end{cases}$$

It is straightforward to see that this NCG is exactly the multicast game.   □

4.3   Equal Consumptions

We know that in the unicast case $\nu = 1 = \nu^*$. We now relax this condition to the following.

$$c(e, S) = c(e, T) \quad \forall e \in E \quad \forall S, T \in \mathfrak{S}(e) . \tag{#}$$

We call (#) the *equal consumption condition*. Under this condition, we have the following theorem. The proof of this is based on the same idea as the ones in the previous section (although it requires some more steps) and has hence been deferred to the full version of this paper.

Theorem 8. *Let $\mathcal{L}$ be standard. Let $\mathcal{I}$ be an instance of selfish multicast routing with latency functions from $\mathcal{L}$. Then $\rho(\mathcal{I}) \leq \alpha(\mathcal{L})$, if we have the equal consumption condition and either a) conservation flows and use l^{edges}, or b) duplication flows and use l^{paths}.*

---

[3]   This requirement can even be relaxed; see [2] for details.

An important observation is the following. Reconsider the example in Fig. 2 with $k = 2$. Then, the equal consumption condition is violated for some links by the minimum possible amount, i.e., by 1. As we have seen in Sect. 3.2, this already may cause an exponential jump in the price of anarchy. If the equal consumption condition was true, we would have a price of anarchy of $O(p)$. But instead we have $\Omega(2^p)$.

5 Future Work

Price of Anarchy. We considered the price of anarchy for several variants of selfish multicast routing and pointed out some cases where the price of anarchy depends on the network topology. For most of these cases, we have lower bounds. We have also upper bounds for polynomial latency functions in two cases. An ongoing effort is to find upper bounds for other variants and more general latency functions and to tighten the existing bounds. We are also interested in establishing bicriteria bounds like in [1, Sec. 3.6].

Taxation. In unicast routing, taxation schemes have shown to be effective in reducing the price of anarchy [6,7]. These schemes make heavy use of the fact that in unicast routing, Nash equilibria are essentially unique, i.e., that all Nash equilibria have the same social cost. Under this assumption, they carry over to some variants of multicast, e.g., to l^{paths} with duplication flows. However, this uniqueness assumption is not always true. For example, consider the example from Thm. 3 with $\nu = k = 2$. Then the flow $\tilde{f}$, which is much better than the equilibrium f, can easily be recognized as an equilibrium as well. This is the reason why the known LP-based taxation schemes fail for this instance. More work will be required to find new schemes.

Another question is how high the taxes will be, e.g., for instances like above with an exponential price of anarchy. Numerical results show that despite different expectations the taxes tend to zero as the price of anarchy grows (exponentially). It would be very interesting to find rigorous proofs for this.

Algorithmic Aspects. Unicast Nash equilibria can be computed efficiently by solving a convex optimization problem. This approach fails for certain variants of multicast, e.g., for l^{paths} and duplication flows. Can new efficient algorithms be established, e.g., based on known numerical methods for variational inequalities?

Other Models. Finally, we intend to leave the Wardrop model, which assumes an infinite number of players, and turn towards a finite number of players, as in the well-known KP-model [8]. For unicast routing on general topologies, approaches in that direction have been made recently [9]. Another aim is a further study of the generalized congestion games mentioned in Remark 2.

6 Summary Table

Table 1. Summary of some of our results. Given are bounds on the worst-case price of anarchy for different variants of multicast and latency functions being polynomials of degree p with non-negative coefficients. A 'u' marks an upper bound, and an 'l' marks a lower bound. The 'top. cond.' refers to the equal consumption condition.

	l^{edges}	l^{paths}	$l^{\text{paths avg}}$	$l^{\max}$
cons.	• $\frac{4}{3}\frac{\nu^2}{\nu^*}$ for $p=1$ (u) • $(p+1)\frac{\nu^{p+1}}{\nu^*}$ (u) • $\Theta(\frac{p}{\ln p})$ (top. cond.) (u & l)	• $\frac{4}{3}$ for $p=1$ (u & l) • $\Theta(\frac{p}{\ln p})$ (u & l)	$\Omega(\frac{R}{r})$ (l)	$\Omega(\frac{R}{r})$ (l)
dupl.	• $\frac{4}{3}$ for $p=1$ (u & l) • $\Theta(\frac{p}{\ln p})$ (u & l)	• $\frac{4}{3}\frac{\nu^2}{\nu^*}$ for $p=1$ (u) • $(p+1)\frac{\nu^{p+1}}{\nu^*}$ (u) • $\Omega(\nu^p)$ (l) • $\Theta(\frac{p}{\ln p})$ (top. cond.) (u & l)	$\Omega(\frac{R}{r})$ (l)	$\Omega(\frac{R}{r})$ (l)

References

1. Roughgarden, T.: Selfish Routing. PhD thesis, Cornell University (2002)
2. Roughgarden, T., Tardos, É.: Bounding the inefficiency of equilibria in nonatomic congestion games. Games and Economic Behavior **47**(2) (2004) 389–403
3. Libman, L., Orda, A.: Atomic resource sharing in noncooperative networks. Telecommunication Systems **17**(4) (2001) 385–409
4. Correa, J.R., Schulz, A.S., Stier Moses, N.E.: Selfish routing in capacitated networks. Mathematics of Operations Research (2004) 961–976
5. Hartman, P., Stampacchia, G.: On some non-linear elliptic differential-functional equations. Acta Mathematica **115** (1966) 271–310
6. Karakostas, G., Kolliopoulos, S.G.: Edge pricing of multicommodity networks for heterogeneous selfish users. In: Proceedings of the 45th Annual IEEE Symposium on Foundations of Computer Science (FOCS'04). (2004)
7. Fleischer, L.K., Jain, K., Mahdian, M.: Tolls for heterogeneous selfish users in multicommodity networks and generalized congestion games. In: Proceedings of the 45th Annual IEEE Symposium on Foundations of Computer Science (FOCS'04). (2004) 277–285
8. Koutsoupias, E., Papadimitriou, C.H.: Worst-case equilibria. In: Proceedings of the 16th International Symposium on Theoretical Aspects of Computer Science (STACS'99). Volume 1563 of Lecture Notes in Computer Science. (1999) 404–413
9. Roughgarden, T.: Selfish routing with atomic players. In: Proceedings of the 15th Annual ACM–SIAM Symposium on Discrete Algorithms (SODA'04). (2004)

Designing a Truthful Mechanism
for a Spanning Arborescence Bicriteria Problem[*]

Davide Bilò[1], Luciano Gualà[1], and Guido Proietti[1,2]

[1] Dipartimento di Informatica, Università di L'Aquila, Italy
[2] Istituto di Analisi dei Sistemi ed Informatica, CNR, Roma, Italy
{davide.bilo, guala, proietti}@di.univaq.it

Abstract. Let a communication network be modelled by a directed graph $G = (V, E)$ of n nodes and m edges, and assume that each edge is owned by a selfish agent, which privately holds a pair of values associated with the edge, namely its *cost* and its *length*. In this paper we analyze the problem of designing a truthful mechanism for computing a *spanning arborescence* of G rooted at a fixed node $r \in V$ having minimum cost (as computed w.r.t. the cost function) among all the spanning arborescences rooted at r which satisfy the following constraint: for each node, the distance from r (as computed w.r.t. the length function) must not exceed a fixed bound associated with the node. First, we prove that the problem is hard to approximate within better than a logarithmic factor, unless NP admits slightly superpolynomial time algorithms. Then, we provide a truthful *single-minded* mechanism for the problem, which guarantees an approximation factor of $(1 + \varepsilon)(n - 1)$, for any $\varepsilon > 0$.

Keywords: Multi-parameter Agents, Algorithmic Mechanism Design, Bicriteria Network Design, Truthful Single-Minded Mechanisms.

1 Introduction

In a non-cooperative game [12] there are several independent agents that have to work together in order to produce an output specification which optimizes a global objective function. However, each agent has its own *valuation* function with respect to a given output, and may speculate in the hope of getting a higher profit. This potentially could lead to economically suboptimal resource allocation and is therefore undesirable. Since the agents are supposed to respond to economic incentives, the main objective of *mechanism design* theory is exactly that of studying how to remunerate the agents in order to let them cooperate with the solving algorithm. A *mechanism* is a pair $\mathcal{M} = \langle g(\cdot), p(\cdot) \rangle$, where $g(\cdot)$ is an algorithm computing a solution, and $p(\cdot)$ specifies the payments provided to the agents. Informally, a mechanism is *truthful* if its payments guarantee that agents are not stimulated to lie. Then, the problem of combining the game theoretic concept of designing a truthful mechanism, and the computational

---

[*] Work partially supported by the Research Project GRID.IT, funded by the Italian Ministry of Education, University and Research.

T. Erlebach (Ed.): CAAN 2006, LNCS 4235, pp. 19–30, 2006.

complexity requirement of designing an efficient algorithm, is exactly the topic of the *algorithmic mechanism design* (AMD) for selfish agents.

In their seminal paper concerned with AMD [11], Nisan and Ronen addressed the classic *shortest path* problem. This problem enjoys the property of being *utilitarian*, in that the quality of any feasible output can be measured by simply summing up all the agents' valuations. For utilitarian problems, there exists a well-known class of truthful mechanisms, i.e., the *Vickrey-Clarke-Groves (VCG) mechanisms* [15,3,5], and therefore the shortest path problem can be solved optimally. Another well-known class of truthful mechanisms is the class of *one-parameter mechanisms* [1]. Informally, a one-parameter mechanism applies to mechanism design problems where the information held by each agent can be expressed by a single parameter. By exploiting the results in [11,1], in a sequel of papers efficient truthful mechanisms have been designed for solving several network design problems [6,7,9,14].

In this paper, given a directed graph $G = (V, E)$ with n nodes and m edges, and with two different functions $c(\cdot)$ and $l(\cdot)$ mapping edges to positive real numbers, we focus on the problem of designing a network topology balancing total cost and distances from a given source node, called *Minimum-cost Arborescence with Bounded Delay* (MABD), which will be defined in more detail in the next section. For this bicriteria problem, we provide the following main results:

- first, we show that the problem of finding a MABD of G has no polynomial time $\kappa \cdot \ln n$-approximation algorithm, for some constant $\kappa > 0$, unless NP admits slightly superpolynomial time algorithms;
- then, under the assumption that each edge of G is owned by a selfish agent, we provide an $\mathcal{O}\left(\frac{mn^4}{\sqrt{1+\varepsilon}-1} \cdot \log \frac{n}{\log(1+\varepsilon)} \cdot \log \frac{n}{\sqrt{1+\varepsilon}-1} \cdot \log \frac{n}{2-\sqrt{1+\varepsilon}}\right)$ time approximated truthful mechanism, with a performance guarantee of $(1 + \varepsilon)(n - 1)$, for any $\varepsilon > 0$.

Notice that the problem we consider cannot be tackled through a one-parameter mechanism, since we allow agents to hold two values, namely the cost and the length of an edge. Rather, as we will see later, our mechanism design problem naturally falls within a generalization of the so-called *single-minded mechanisms* (see Section 2.2 for a formal definition). These were originally introduced by Lehmann *et al.* [10] in the framework of *combinatorial auctions*, and then were extended by Briest *et al.* [2] to handle a wider class of optimization problems. In particular, in [2] the authors sketched the existence of approximate truthful mechanisms for two classic bicriteria network design problems, namely the *constrained shortest path problem* (see Section 2.1 for a definition of the problem) and the *constrained minimum spanning tree problem*.

The paper is organized as follows: in Section 2 we give the definition of the problem and recall some basic notions from the mechanism design theory; in Section 3 we present an inapproximability results for the MABD problem, while in Section 4 we provide an approximate monotone algorithm for it; finally, in Section 5 we design an approximate truthful mechanism for the MABD problem.

2   Preliminaries

Let $G = (V, E)$ be a directed graph, with n nodes and m edges, and with two different functions $c(e)$ and $l(e)$ mapping edges to positive real numbers. We will call $c(e)$ the *cost* of e, and $l(e)$ the *length* of e. A graph $H = (V(H), E(H))$ is called a *subgraph* of G if $V(H) \subseteq V$ and $E(H) \subseteq E$. If $V(H) = V$ then H is called a *spanning subgraph* of G. A (directed) *simple path* P (or a *path* for short) from v_1 to v_k in G is a subgraph with $V(P) = \{v_1, \ldots, v_k \mid v_i \neq v_j$ for $i \neq j\}$ and $E(P) = \{e_i = (v_i, v_{i+1}) \mid 1 \leq i < k\}$, and it is denoted by $(v_1, \ldots, v_k)$. Given a node $r \in V$, a *spanning arborescence* of G is a spanning tree of G rooted at r and having all the edges oriented towards the leaves. Given a source node r and a destination node s, a path in G from r to s is a *shortest path*, say $P_G(r, s)$, if the sum of its edge lengths (called *distance* between r and s in G, and denoted by $d_G(r, s)$) is minimum. We define the *total cost* of a spanning subgraph H of G as $c(H) = \sum_{e \in E(H)} c(e)$. Finally, for each edge $e \in E$, we denote by $G - e$ the graph $(V, E \setminus \{e\})$.

2.1   The MABD Problem

We are interested in finding a spanning tree structure balancing total cost and distances from a given source node. Several problems can be defined according to this goal. In this paper, we will focus on the following problem: Given a source node $r \in V$, and given a positive real budget value L_v for each node $v \in V$, the *Minimum-cost Arborescence with Bounded Delay* (MABD) problem asks for computing a cheapest (i.e., of minimum total cost) spanning arborescence T of G rooted at r which satisfies the constraint that for each node $v \in V$, the distance from r to v in T is at most L_v, i.e., $d_T(r, v) \leq L_v$.

A problem related to the MABD one is the *constrained shortest path problem*, in which one looks for a cheapest path from a source node r to a destination node s among all paths of length L or less. Such problem is known to be (weakly) NP-hard, and it admits a pseudo-polynomial time algorithm, which is transformed by scaling techniques to a fully polynomial time approximation scheme (FPTAS) for the problem [13,8]. Moreover, in [2] it is shown how to make the algorithm in [13] *monotone*, which is needed to design a truthful mechanism for the same problem (see next section for the definition of monotonicity).

2.2   A Non-cooperative Setting: Monotonicity and Truthfulness

Algorithmic mechanism design deals with algorithmic problems in a non-cooperative setting, in which part of the input is owned by selfish agents. As such agents may lie about their parts of input, they are capable of manipulating the algorithm. The main task of the mechanism design theory is the study of how to pay the agents in order to convince them to cooperate with the algorithm. We will deal with the case in which each agent controls a single link of a communication network modelled by a directed graph $G = (V, E)$. We provide a simplified formalization below, and we refer the interested reader to [11,1].

For an edge e of G owned by a selfish agent a_e, we denote by $\hat{c}(e)$ and $\hat{l}(e)$ the private information held by a_e w.r.t. the cost and the length of e, respectively. Intuitively, $\hat{c}(e)$ and $\hat{l}(e)$ represents the true cost and the true delay of forwarding a message through the link e, respectively. We call $t_e = \langle \hat{c}(e), \hat{l}(e) \rangle$ the (private) *type* of the agent a_e. Each agent has to declare a (public) *bid* $b_e = \langle c(e), l(e) \rangle$ to the mechanism. We will denote by t the vector of private types, and by b the vector of bids.

For a given optimization problem defined on G, let $\mathcal{F}$ denote the corresponding set of feasible solutions. For each feasible solution $x \in \mathcal{F}$, some measure function $\mu(x, t)$ is defined, which depends on the true types. A *mechanism* is a pair $\mathcal{M} = \langle g(b), p(b) \rangle$, where $g(b)$ is an algorithm that, given agents' bids, computes a solution, and $p(b)$ is a scheme which describes the payments provided to the agents. A mechanism has a runtime of $\mathcal{O}(f(n))$ if $g(\cdot)$ and $p(\cdot)$ are computable in $\mathcal{O}(f(n))$ time. For each solution x and for each agent a_e, it is defined a public *valuation* function $\nu_e(x, t_e)$, which represents the expense incurred by a_e in x. The *utility* function $u_e(g(b), t_e)$ of an agent is defined as the difference between the payment provided by the mechanism and its valuation w.r.t. the computed solution. Each agent tries to maximize its utility, while an *exact* mechanism aims to compute a solution which optimizes $\mu(x, t)$, but of course it does not know t directly. Similarly, if we denote by $\varepsilon(n)$ a positive real function of the input size n, an $\varepsilon(n)$-*approximation* mechanism returns a solution $g(b)$ whose measure comes within a factor $\varepsilon(n)$ from the optimum. In a *truthful* mechanism the tension between the agents and the system is resolved, since each agent maximizes its utility when it declares its type, regardless of what the other agents do.

We are going to deal with a special type of agents called *single-minded*, as defined by Lehmann et al. [10] in the framework of *combinatorial auctions*, and lately extended by Briest et al. [2] in order to handle a wider class of optimization problems. Informally, a single-minded agent is able to offer only a single contribution to a solution (possibly defined by multiple parameters). Concerning our problem, the algorithm of the mechanism computes a weighted arborescence T, and the single contribution of an agent is exactly the two-parameter edge owned by the agent. The definition of the valuation function of our single-minded agent a_e thus reduces to that given in [2]:[1]

$$\nu_e(T, t_e) = \begin{cases} \hat{c}(e) & \text{if } e \in E(T) \text{ and } l(e) \geq \hat{l}(e); \\ +\infty & \text{if } e \in E(T) \text{ and } l(e) < \hat{l}(e); \\ 0 & \text{otherwise.} \end{cases}$$

---

[1] To be more precise, the authors in [2] gave the formal definition of a generalized single-minded agent by focusing on maximization problems, which is slightly different from the one concerning minimization problems. However, they also considered some minimization problems, for which it is easy to see that the corresponding valuation functions are implicitly defined exactly as in our case (see for example the constrained shortest path problem discussed in [2]).

The idea behind the above valuation function is the following: a selected agent can forward a message in time strictly less than $\hat{l}(e)$ only by incurring a very large cost, while otherwise its cost is equal to $\hat{c}(e)$.

Notice that our problem enjoys the property of being *utilitarian*, since our measure function satisfies $\mu(x,t) = \sum_{e\in E} \nu_e(x,t_e)$. For utilitarian problems with single-minded agents, a sufficient condition for truthfulness is given by a particular property of the mechanism algorithm. Let b_{-e} denote the vector of all bids besides b_e, while let (b_{-e}, b_e) denote the vector b.

Definition 1 ([10,2]). *An algorithm is said to be* monotone *if for each bid vector b, whenever it selects an edge e in b, it still selects e in $b' = (b_{-e}, \langle c'(e), l'(e)\rangle)$, for any $c'(e) \leq c(e)$ and $l'(e) \leq l(e)$; or, equivalently, if for each non-selected edge e in b, e is still non-selected in $b' = (b_{-e}, \langle c'(e), l'(e)\rangle)$, for any $c'(e) \geq c(e)$ and $l'(e) \geq l(e)$.*

If we fix b_{-e} and $l(e)$, a monotone algorithm defines a threshold value θ_e such that if a_e bids no more than θ_e, then e will be selected, while if a_e bids above θ_e, then e will not be selected. Therefore, the following general result holds:

Theorem 1 ([10,2]). *Let $\mathcal{M} = \langle g(\cdot), p(\cdot)\rangle$ be a mechanism for some utilitarian problem with single-minded agents. Then $\mathcal{M}$ is truthful if and only if $g(\cdot)$ is monotone and the payment for each agent e is defined as its threshold value θ_e if it owns a selected edge, and 0 otherwise.*

3   The Hardness Result

Our non-approximability result for the MABD problem is obtained by a reduction (preserving the approximation) from the *Set Cover Problem* (SCP). An instance $I = \langle O, \mathcal{S}\rangle$ for the SCP consists of a set $O = \{o_1, \ldots, o_h\}$ of h objects, and a set $\mathcal{S} = \{S_1, \ldots, S_\ell\}$ of ℓ subsets of O. The objective is to find a minimum-size collection of subsets in $\mathcal{S}$ whose union is O. In [4] it is shown that SCP cannot be approximated within $(1 - o(1))\ln h$, unless $\mathsf{NP} \subseteq \mathsf{DTIME}(h^{\mathcal{O}(\log \log h)})$. The same result holds even for the case $\ell \leq h$ [4]. The following holds:

Lemma 1. *Let $L > 0$ be an integer. Then, the MABD problem has no polynomial time approximation algorithm with a performance guarantee better than $(1 - o(1))\ln \frac{n-2}{L}$, where $n \geq L + 2$, unless $\mathsf{NP} \subseteq \mathsf{DTIME}\left(\left(\frac{n}{L}\right)^{\mathcal{O}(\log \log \frac{n}{L})}\right)$, even for the unit length case and when the maximum delay is L for every edge.*

Proof. Let $I = \langle O, \mathcal{S}\rangle$ be an instance for the SCP with $\ell \leq h$, and h such that $L \leq h^k$ for some integer k. From I we build an instance $\mathcal{I} = \langle D, c, l, r, L\rangle$ for the MABD problem in the following way (see Figure 1). The node set of G is defined as follows:

- a node r, which is the source;
- $\lfloor L/2 \rfloor$ nodes $s_1^i, \ldots, s_{\lfloor L/2\rfloor}^i$, for each $S_i \in \mathcal{S}$;

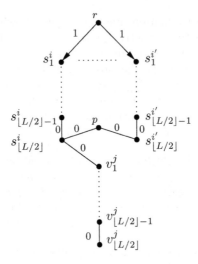

Fig. 1. The reduction of Lemma 1 when L is odd. Undirected edges denote the two corresponding directed ones. Note that here o_j belongs to S_i, but o_j does not belong to $S_{i'}$.

- $\lfloor L/2 \rfloor$ nodes $v_1^j, \ldots, v_{\lfloor L/2 \rfloor}^j$, for each $o_j \in O$;
- a node p only if L is odd.

The edge set of G is defined as follows:

- an edge (r, s_1^i) of cost 1, for each $S_i \in \mathcal{S}$;
- $2(\lfloor L/2 \rfloor - 1)$ edges (s_k^i, s_{k+1}^i) and (s_{k+1}^i, s_k^i), $k = 1, \ldots, (\lfloor L/2 \rfloor - 1)$, of cost 0, for each $S_i \in \mathcal{S}$;
- $2(\lfloor L/2 \rfloor - 1)$ edges (v_k^j, v_{k+1}^j) and (v_{k+1}^j, v_k^j), $k = 1, \ldots, (\lfloor L/2 \rfloor - 1)$, of cost 0, for each $o_j \in O$;
- two edges $(s_{\lfloor L/2 \rfloor}^i, v_1^j)$ and $(v_1^j, s_{\lfloor L/2 \rfloor}^i)$ of cost 0, for each S_i and o_j such that $o_j \in S_i$;
- two edges $(s_{\lfloor L/2 \rfloor}^i, s_{\lfloor L/2 \rfloor}^j)$ and $(s_{\lfloor L/2 \rfloor}^j, s_{\lfloor L/2 \rfloor}^i)$ of cost 0, for each $S_i, S_j \in \mathcal{S}$, if L is even, and four edges $(s_{\lfloor L/2 \rfloor}^i, p), (s_{\lfloor L/2 \rfloor}^j, p), (p, s_{\lfloor L/2 \rfloor}^i)$ and $(p, s_{\lfloor L/2 \rfloor}^j)$ of cost 0, for each $S_i, S_j \in \mathcal{S}$, if L is odd.

Every edge has length 1 and the distance requirement is L for each node. We claim that G has an arborescence with maximum delay L of cost k if and only if the original instance of the SCP has a solution of size k.

It is easy to see that a solution $\mathcal{C}$ for the SCP instance provides a solution for $\mathcal{I}$ with the same total cost. Indeed, the feasible solution is given by the arborescence T defined as follows:

- begin with $H = (V(G), \emptyset)$
- for each set $S_i \in \mathcal{C}$, H contains the path $\left(r, s_1^i, s_2^i, \ldots, s_{\lfloor L/2 \rfloor}^i\right)$, and all the paths $\left(s_{\lfloor L/2 \rfloor}^i, v_1^j, v_2^j, \ldots, v_{\lfloor L/2 \rfloor}^j\right)$, for each $o_j \in S_i$;

– for each set $S_j \notin \mathcal{C}$, H contains either the path $\left(s^i_{\lfloor L/2\rfloor}, s^j_{\lfloor L/2\rfloor}, s^j_{\lfloor L/2\rfloor -1}, \ldots, s^j_1\right)$ if L is even, or the path $\left(s^i_{\lfloor L/2\rfloor}, p, s^j_{\lfloor L/2\rfloor}, s^j_{\lfloor L/2\rfloor -1}, \ldots, s^j_1\right)$ if L is odd, where $S_i \in \mathcal{C}$.

– Let T be a *shortest path tree*[2] of H.

Now, let T be a solution for $\mathcal{I}$ of cost $k \leq \ell$ (notice that such a solution always exists). Since T has maximum delay L, it is not hard to see that for each $v^j_{\lfloor L/2\rfloor}$, T must have a path $\left(r, s^i_1, s^i_2, \ldots, s^i_{\lfloor L/2\rfloor}, v^j_1, v^j_2, \ldots, v^j_{\lfloor L/2\rfloor}\right)$, for some $S_i \in \mathcal{S}$ such that $o_j \in S_i$. Hence, $\mathcal{C} = \{S_i \in \mathcal{S} \mid (r, s^i_1) \in E(T)\}$ is a solution for $\mathcal{I}$ of size no more than the total cost of T. Since $\ell \leq h$, the number of nodes n is at most $2\lfloor L/2\rfloor h + 2$. The claim now follows from the inapproximability result for the SCP proved in [4]. $\qquad\square$

4   An Approximate Monotone Algorithm

The algorithm computes a set $\Pi = \{\pi_v : v \in V \setminus \{r\}\}$ of "light" paths, meaning that π_v is a $(1 + \xi)$-approximation of a cheapest path of length at most L_v, for any $\xi > 0$. Then, it considers the subgraph H of G consisting of the union of all these paths. As H may have cycles, then the algorithm needs to remove edges from H so that a feasible arborescence T of G is returned.

We point out that the removal step (Lines 5–7) must be defined in order to guarantee both the monotonicity property and the feasibility of the solution. For instance, computing a shortest path tree of H yields to a non-monotone algorithm. In the pseudo-code given below, we use the following notation. For each node v, we define a measure $\delta(v)$ as follows:

$$\delta(v) = \min_{\pi_u:\, v \in V(\pi_u)} \{l(\pi_u[r, v])\},$$

where by $\pi_u[r, v]$ we denote the subpath of π_u going from r to v, while $l(\pi_u[r, v]) = \sum_{e \in E(\pi_u[r,v])} l(e)$. Moreover for each node v, we define the *winning path* π^*_v in Π as[3]

$$\pi^*_v = \arg \min_{\pi_u \in \Pi:\, v \in V(\pi_u)} \{l(\pi_u[r, v])\},$$

and the *winning edge* e^*_v to be the edge (u, v) of π^*_v.

Finally, as we are interested in monotone algorithms, we use the monotone FPTAS for the constrained shortest path problem provided in [2] for computing each path π_v in Line 2. We denote by $\mathcal{A}$ such algorithm.

---

[2] A *shortest path tree* is a tree consisting of the union of all the shortest paths from r to every other node.

[3] Notice that the winning path may not be unique. Since we need our algorithm to be monotone, we can say the winning path is unique as we can assume that nodes are labelled with different integers, and hence in a case of a tie, the winning path is the one whose ending node has the lowest label. Notice that this selection rule has the nice property of being monotone. As a consequences, we may assume that the winning edge is unique as well.

Algorithm 1

Input: $G = (V, E, c, l)$, $r \in V$, $L = (L_{v_1}, \ldots, L_{v_{n-1}})$, $L_{v_i} \in \mathbb{R}^+$, $\xi > 0$.
Output: an arborescence T of G rooted at r.
1: **for** each $v \in V$ **do**
2:    find a $(1 + \xi)$-apx π_v of a cheapest path from r to v with $l(\pi_v) \le L_v$;
3: **end for**
4: Let T be the digraph made up of the union of all edges in π_v, $\forall v \in V$
5: **for** each $v \in V \setminus \{r\}$ **do**
6:    remove all edges in T entering in v but the winning one e_v^*
7: **end for**
8: **return** T

Lemma 2. *Algorithm 1 returns an arborescence of G.*

Proof. At the end of the algorithm each node in T but r has in-degree 1. Hence, to prove that T is an arborescence rooted at r, it remains to show that there is a path in T from r to every other node v.

Let $v_0 = r$, v_1, $\ldots$, v_{n-1} be the nodes ordered by measure $\delta(\cdot)$. The proof is by induction on i. The basic case is $i = 0$, which is trivial. Let $i > 0$, and suppose that every v_j is reachable in T from r, $j < i$. Let v_i be the $(i+1)$-th node in the ordering. Let (v, v_i) be the winning edge for v_i. Since $l(\pi_{v_i}^*[r, v]) < l(\pi_{v_i}^*[r, v_i])$, and since $\delta(v) \le l(\pi_{v_i}^*[r, v])$, it follows that v comes first than v_i in the ordering. Then, by the inductive hypothesis, v is reachable, and since the winning edge is (v, v_i), then v_i is also reachable, from which the claim follows. $\square$

Lemma 3. *Algorithm 1 returns a feasible solution.*

Proof. We prove that, at the end of the algorithm, $d_T(r, v) \le \delta(v), \forall v \in V$. The lemma will follow from the fact that $\delta(v) \le L_v$ for every v.

The proof is by induction on the number of edges of the path $P_T(r, v)$. The basic case is $v = r$, which is trivial. Now, consider the case in which $P_T(r, v)$ consists of $k > 0$ edges, and assume that for every node u for which $P_T(r, u)$ is made up of $h < k$ edges, it holds $d_T(r, u) \le \delta(u)$. Let (v', v) be the winning edge for v. Since $P_T(r, v')$ has less than k edges, then by the inductive hypothesis, we have $d_T(r, v') \le \delta(v')$, and thus

$$d_T(r, v) = d_T(r, v') + l(v', v) \le \delta(v') + l(v', v) \le l(\pi_v^*[r, v']) + l(v', v) = \delta(v).$$

$\square$

Lemma 4. *Algorithm 1 returns a $(1 + \xi)(n - 1)$-approximated solution for the MABD problem.*

Proof. Let $\mathcal{T}$ be an optimal solution for the problem. For each node $v \in V \setminus \{r\}$, the path $P_{\mathcal{T}}(r, v)$ has length at most L_v. Since each π_v is a $(1+\xi)$-approximation solution of an optimal low cost path from r to v of length at most L_v, we have

$$c(\pi_v) \le (1 + \xi) c(P_{\mathcal{T}}(r, v)).$$

Hence

$$c(T) \leq \sum_{v \in V \setminus \{r\}} c(\pi_v) \leq \sum_{v \in V \setminus \{r\}} (1 + \xi)\, c(P_T(r, v)) \leq (1 + \xi)(n - 1)\, c(T).$$

□

Lemma 5. *Algorithm 1 is monotone.*

Proof. Let b be the vector of agents' bids, and let $e = (v', v)$ be a selected edge in b. Moreover, let $c'(e) \leq c(e)$ and $l'(e) \leq l(e)$. We have to show that the algorithm will still select the edge e in $b' = (b_{-e}, \langle c'(e), l'(e) \rangle)$.

Let v be the node for which e is the winning edge. Let $\Pi(b)$ be the set of paths selected by the algorithm in Lines 1–3 when the input is b, and let $\Pi_{e'}(b) \subseteq \Pi(b)$ be the subset of paths passing through e', where e' is an edge entering in v. From the monotonicity of algorithm $\mathcal{A}$, we have that $\Pi_e(b') \supseteq \Pi_e(b)$ and $\Pi_{e'}(b') \subseteq \Pi_{e'}(b)$, for any edge $e' \neq e$ entering in v. This shows that the winning path for the node v in b' must belong to $\Pi_e(b')$, and thus e will be still selected.   □

Theorem 2. *For any $\xi > 0$, there exists a $(1 + \xi)(n - 1)$-approximate monotone algorithm for the MABD problem, running in $\mathcal{O}\left(\frac{mn^3}{\xi} \log \frac{n^2}{\xi} \log \frac{n}{1 - \xi}\right)$ time.*

Proof. Approximation ratio and monotonicity follows from Lemma 4 and from Lemma 5. Concerning the time complexity, $\mathcal{A}$ runs in $\mathcal{O}\left(\frac{mn^2}{\xi} \log \frac{n^2}{\xi} \log \frac{n}{1 - \xi}\right)$ time [2], while the rest is bounded by $\mathcal{O}(n^2)$. The claim follows.   □

5   An Approximate Truthful Mechanism

Our goal is to design a mechanism that satisfies the following conditions: (i) the selection rule defined by $g(\cdot)$ is monotone; (ii) the solution returned by $g(\cdot)$ is a $(1 + \varepsilon)(n - 1)$-approximation of the optimal solution, for any $\varepsilon > 0$; and (iii) $g(\cdot)$ and $p(\cdot)$ are computable in polynomial time. These three questions are addressed in the following subsections.

5.1   The Algorithm of the Mechanism

Our main idea is to apply rounding on the costs as this simplifies the computation of the payments. Given the vector of bids b and $\varepsilon > 0$, we round the costs of the edges as follows.

For each $e \in E$, the *rounded cost* $c_R(e)$ of e, is defined as $c_R(e) = (1 + \xi)^k$, where $k \in \mathbb{Z}$ is the (unique) integer such that $(1 + \xi)^{k-1} < c(e) \leq (1 + \xi)^k$, and $\xi = \sqrt{1 + \varepsilon} - 1$.

Next, we use Algorithm 1 to compute a $(1 + \xi)$-approximation of the optimum solution for the MABD w.r.t. the rounded costs. The following holds:

Lemma 6. *The algorithm of the mechanism defined above is monotone.*

Proof. For any edge e, $c(e) \leq c'(e)$ implies $c_R(e) \leq c'_R(e)$. Now the claim follows from the monotonicity of Algorithm 1 (see Lemma 5). $\qquad\square$

Let b be the vector of bids, and let $\mathcal{T}$ be an optimum solution of the original instance (without rounded costs). The following lemma holds:

Lemma 7. *Let T be the solution returned by the algorithm on the instance with rounded costs. Then $c(T) \leq (1 + \varepsilon)(n - 1)\, c(\mathcal{T})$.*

Proof. Let $\mathcal{T}_R$ be an optimum solution of the instance with rounded costs. From the optimality of $\mathcal{T}_R$, we have that $c_R(\mathcal{T}_R) \leq c_R(\mathcal{T})$. By using Lemma 4, we obtain:

$$c(T) \leq c_R(T) \leq (1 + \xi)(n - 1)\, c_R(\mathcal{T}_R) \leq (1 + \xi)(n - 1)\, c_R(\mathcal{T})$$
$$\leq (1 + \xi)^2 (n - 1)\, c(\mathcal{T}) = (1 + \varepsilon)(n - 1)\, c(\mathcal{T}),$$

where the last inequality holds because $c_R(e) \leq (1 + \xi)\, c(e)$, for any edge e. $\quad\square$

5.2   The Payment Scheme

In this subsection we show how to compute the payments. We remind that from Theorem 1, the payment for the edge e is defined as the threshold value (w.r.t. the algorithm of our mechanism) θ_e for each winning edge, and 0 otherwise.

Let $e \in \pi_u$, for some u. We define the *threshold* $\theta(e, u)$ *of* e w.r.t. π_u as the minimum value such that if $c(e) > \theta(e, u)$ then e exits from π_u. Note that $\theta(e, u)$ must exist from the monotonicity of $\mathcal{A}$.[4] Consider the interval $X = [c_R(e), c_R(E)]$, and let k be the integer such that $c_R(e) = (1 + \xi)^k$. We define a set of *points* in X as follows:

$$x_i = (1 + \xi)^{i+k} \quad i = 0, 1, \dots.$$

Then it must exist an index j such that if $c_R(e) = x_j$, then e is selected in π_u, while e does not belong to π_u if $c(e) = x_{j+1}$. It is clear that $\theta(e, u) = x_j$. We compute j by performing a binary search on the set of indexes and repeatedly running $\mathcal{A}$ to check whether e is selected.

Now we show how to compute the payments θ_e for each winning edge e. Let e be a winning edge entering in v, and let $\Pi_e \subseteq \Pi$ be the set of paths in Π containing e. We denote by π_v^{-e} the winning path for v in $\Pi \setminus \Pi_e$. Let $\pi_{u_1}, \dots, \pi_{u_h}$ be all the paths in Π_e ordered in a non-decreasing way w.r.t. the threshold values $\theta(e, u_i)$. For the sake of clarity, we will denote by l_i the length of the path $\pi_{u_i}[r, v]$ and by l^{-e} the length of the path $\pi_v^{-e}[r, v]$.

Moreover, let $\pi_{u_i}^{-e}$ be the constrained shortest path from r to u_i computed by $\mathcal{A}$ in $G - e$ with rounded costs. Finally, we define l_i^{-e} as follows:

$$l_i^{-e} = \begin{cases} l(\pi_{u_i}^{-e}[r, v]) & \text{if } \pi_{u_i}^{-e} \text{ passes through } v; \\ \infty & \text{otherwise.} \end{cases}$$

---

[4] As usual, we assume that there always exists an alternative feasible path from r to u in $G - e$, otherwise e may declare any value.

It is not hard to see that $\theta_e = \theta(e, u_i)$, where i is the minimum index such that

$$\min\{l_{i+1}, \ldots, l_h\} > \min\{l^{-e}, l_1^{-e}, \ldots, l_i^{-e}\}.$$

5.3   Mechanism Runtime

In this subsection we prove that our mechanism can be computed in polynomial time, and we finally provide our main result.

Lemma 8. *All $\theta(e, u_i)$ values can be computed in polynomial time.*

Proof. We compute each $\theta(e, u_i)$ by performing a binary search. Since the number of points is bounded by

$$|\log_{1+\xi}(c_R(e))| + |\log_{1+\xi}(c_R(E))| = \frac{|\log c_R(e)|}{\log(1+\xi)} + \frac{|\log c_R(E)|}{\log(1+\xi)}$$

$$\leq \frac{2\max\{|\log c_R(e)|, |\log c_R(E)|\}}{\log(1+\xi)},$$

a binary search takes

$$\mathcal{O}\big(\log(\max\{\log_{1+\xi} c_R(e), \log_{1+\xi} c_R(E)\})\big) = \mathcal{O}\left(\log \frac{n}{\log(1+\xi)}\right),$$

since $\max\{\log c_R(e), \log c_R(E)\}$ is polynomial in n, for any reasonable encoding. The time complexity of $\mathcal{A}$ is $\mathcal{O}\left(\frac{mn^2}{\xi} \log \frac{n^2}{\xi} \log \frac{n}{1-\xi}\right)$ [2]. Since we have to perform $\mathcal{O}(n^2)$ binary searches, one for each pair (e, π_u), such that e is a winning edge and $e \in \pi_u$, the overall time complexity is

$$\mathcal{O}\left(\frac{mn^4}{\sqrt{1+\varepsilon}-1} \cdot \log \frac{n}{\log(1+\varepsilon)} \cdot \log \frac{n}{\sqrt{1+\varepsilon}-1} \cdot \log \frac{n}{2-\sqrt{1+\varepsilon}}\right)$$

as $\xi = \sqrt{1+\varepsilon} - 1$. $\qquad\square$

Lemma 9. *All θ_e values can be computed in polynomial time.*

Proof. We repeatedly use $\mathcal{A}$ for computing every $\pi_{u_i}^{-e}$ (and every l_i^{-e}) for each winning edge e and for each u_i. This takes $\mathcal{O}\left(\frac{mn^4}{\sqrt{1+\varepsilon}-1} \cdot \log \frac{n}{\sqrt{1+\varepsilon}-1} \cdot \log \frac{n}{2-\sqrt{1+\varepsilon}}\right)$ time. It is easy to see that all other tasks can be accomplished in $\mathcal{O}(n^2 \log n)$ time. By Lemma 8 the claim follows. $\qquad\square$

From Lemmas 7–9 and from Theorems 1 and 2, we can finally state the following:

Theorem 3. *Given any $\varepsilon > 0$, there exists a $(1 + \varepsilon)(n - 1)$-approximate truthful mechanism for the MABD problem, running in $\mathcal{O}\left(\frac{mn^4}{\sqrt{1+\varepsilon}-1} \cdot \log \frac{n}{\log(1+\varepsilon)} \cdot \log \frac{n}{\sqrt{1+\varepsilon}-1} \cdot \log \frac{n}{2-\sqrt{1+\varepsilon}}\right)$ time.* $\qquad\square$

References

1. A. Archer and É. Tardos, Truthful mechanisms for one-parameter agents, *Proc. 42nd IEEE Symp. on Foundations of Computer Science (FOCS'01)*, 482–491, 2001.
2. P. Briest, P. Krysta, B. Vöcking, Approximation techniques for utilitarian mechanism design, *Proc. 37th Ann. ACM Symp. on Theory of Computing (STOC'05)*, 39–48, 2005.
3. E. Clarke, Multipart pricing of public goods, *Public Choice*, 8:17–33, 1971.
4. U. Feige, A threshold of $\ln n$ for approximating set cover, *J. ACM*, 45(4):634-652, 1998.
5. T. Groves, Incentives in teams, *Econometrica*, 41(4):617–631, 1973.
6. L. Gualà and G. Proietti, A truthful $(2\text{-}2/k)$-approximation mechanism for the Steiner tree problem with k terminals, *Proc. 11th Int. Computing and Combinatorics Conference (COCOON'05)*, Vol. 3595 of Lecture Notes in Computer Science, Springer-Verlag, 390–400.
7. L. Gualà and G. Proietti, A truthful efficient truthful mechanisms for the single-source shortest paths tree problem, *Proc. 11th Int. Euro-Par Conf. (Euro-Par'05)*, Vol. 3648 of Lecture Notes in Computer Science, Springer-Verlag, 941-951.
8. R. Hassin, Approximation schemes for restricted sortest path problems, *Math. Oper. Res.*, 17(1):36–42, 1992.
9. J. Hershberger and S. Suri, Vickrey prices and shortest paths: what is an edge worth?, *Proc. 42nd IEEE Symp. on Foundations of Computer Science (FOCS'01)*, 252–259.
10. D. Lehmann, L. O'Callaghan, and Y. Shoham, Truth revelation in approximately efficient combinatorial auctions, *ACM conference of Electronic Commerce (EC'99)*, 96–102, 1999.
11. N. Nisan and A. Ronen, Algorithmic mechanism design, *Games and Economic Behaviour* 35:166–196, 2001.
12. M.J. Osborne and A. Rubinstein, *A Course in Game Theory*, MIT Press, 1994.
13. C.A. Phillips, The network inhibition problem, *Proc. 25th Ann. ACM Symp. on Theory of Computing (STOC'93)*, 776–785, 1993.
14. G. Proietti and P. Widmayer, A truthful mechanism for the non-utilitarian minimum radius spanning tree problem, *Proc. 17th ACM Symp. on Parallelism in Algorithms and Architectures (SPAA'05)*, 195–202, 2005.
15. W. Vickrey, Counterspeculation, auctions and competitive sealed tenders, *J. of Finance*, 16:8–37, 1961.

On the Topologies of Local Minimum Spanning Trees

(Extended Abstract)

P.F. Cortese[1], G. Di Battista[1], F. Frati[1], L. Grilli[2],
K.A. Lehmann[3], G. Liotta[2], M. Patrignani[1], I.G. Tollis[4], and F. Trotta[2]

[1] Dipartimento di Informatica e Automazione, Università Roma Tre, Italy
[2] Dipartimento di Ingegneria Elettronica e dell'Informazione, Università di Perugia, Italy
[3] Wilhelm-Schickard Institut für Informatik, University of Tübingen, Germany
[4] Inst. of Computer Science, Foundation for Research and Technology Hellas-FORTH, Greece

Abstract. This paper is devoted to study the combinatorial properties of Local Minimum Spanning Trees (LMSTs), a geometric structure that is attracting increasing research interest in the wireless sensor networks community. Namely, we study which topologies are allowed for a sensor network that uses, for supporting connectivity, a local minimum spanning tree approach. First, we refine the current definition of LMST realizability, focusing on the role of the power of transmission (i.e., of the radius of the covered area). Second, we show simple planar, connected, and triangle-free graphs with maximum degree 3 that cannot be represented as an LMST. Third, we present several families of graphs that can be represented as LMSTs. Then, we show a relationship between planar graphs and their representability as LMSTs based on homeomorphism. Finally, we show that the general problem of determining whether a graph is LMST representable is NP-hard.

1 Introduction

A wireless sensor network consists of a collection of geographically distributed vertices (sensors) that communicate with one another through a wireless medium. A wireless sensor network can be modeled as a set of points in the plane where any sensor s can communicate directly with each other sensor that is inside a region surrounding s that represents its power range. Such a region is often modeled as a unit circle. This model gives rise to a geometric graph called *unit disk graph* (UDG), where each sensor s is a vertex of the graph and there is an edge connecting s to another sensor t if and only if t is within the power range of s, i.e., t is in the unit circle centered at s. However, depending on the location of the sensors, the UDG may be too dense for the limited memory of the sensors in the network; also, in order to reduce energy consumption, it is desirable that each sensor communicates directly with only a few of the sensors that are within its range. Indeed, topology control and management, i.e. how to maintain the sensor network's connectivity while consuming the limited transmission power, has emerged to be one of the important issues in wireless networks (see, e.g., [10]).

An increasing number of topology control algorithms have thus been presented in the literature that are based on geometric graphs which are sparser than the UDG and have

T. Erlebach (Ed.): CAAN 2006, LNCS 4235, pp. 31–44, 2006.

a number of additional interesting properties, such as having small vertex degree, or being good spanners, or having the longest edge relatively short, or being efficiently computable in a distributed manner where each sensor s only knows the location of those sensors that can be reached with a few hops from s. A limited list of these structures includes *k-localized Delaunay triangulations* (see, e.g., [16]), *local minimum spanning trees* (see, e.g., [14,3]), and *partial Delaunay triangulations* (see, e.g., [18]). The reader is also referred to the survey of Li [15] for definitions and more references. We recall here that the knowledge of the combinatorial properties of the communication network is a basic requirement for the design of efficient localized routing algorithms (see, e.g., [1,12,17]).

We also recall that the investigation of the combinatorial properties of geometric graphs has a long tradition in the computational geometry and graph drawing communities (see, e.g., [20,6,9,13,21]).

The present paper is devoted to study the combinatorial properties of local minimum spanning trees, a model that is attracting increasing research interest. We say that a graph G is LMST drawable if it admits a straight-line drawing that is the local minimum spanning tree of the set of its vertices. In [3,14] it has been proved that the maximum vertex degree of an LMST drawable graph is at most 6, that an LMST drawable graph is always planar, and that it is also connected if the underlying UDG is connected. This paper shades more light on the family of LMST drawable graphs:

- We refine the definition of LMST drawable graph, focusing on the role of the radius in the drawability of a graph.
- We show a simple planar, connected, and triangle-free graph with maximum degree 3 that is not LMST drawable.
- We give several positive results on the LMST drawability of graphs. Namely, we show that graphs that are homeomorphic to the orthogonal (hexagonal) grid, subgraphs of the orthogonal grid, and outerplanar graphs with maximum degree 4 and whose dual is a path, are LMST drawable.
- We present a relationship between planar graphs and their drawability based on homeomorphism that, as far as we know, is a novel concept. Namely, we show that every n-vertex planar graph with maximum degree 5 has a homeomorphic planar graph that is LMST drawable. Further, we show that for every n-vertex planar (outerplanar) graph G with maximum degree 4 there exists a planar graph with $O(n^2)$ $(O(n))$ vertices homeomorphic to G that is LMST drawable. Finally, we show that there exists an n-vertex planar graph that does not have any homeomorphic planar LMST drawable graph with less than $\Omega(n)$ extra vertices.
- We show that the decision whether a graph is LMST$^+$ drawable is *NP-hard* (the definition of LMST$^+$ is given in Section 2).

The rest of this paper is organized as follows. In Section 2 we introduce basic terminology. In Section 3 we show several families of graphs that are LMST drawable and a planar, connected, and triangle-free graph with maximum degree 3 that is not LMST drawable. In Section 4 we study the interplay between graph homeomorphism and LMST drawability. The proof that the general LMST$^+$ drawability problem is NP-hard is given in Section 5. Finally, Section 6 contains conclusions and open problems.

2   Preliminaries

We assume familiarity with basic graph drawing concepts [5]. A drawing of a graph is *planar* if no two segments cross. A *grid drawing* is such that all its vertices have integer coordinates. A *straight-line drawing* is such that all edges are rectilinear segments. A *polyline drawing* is such that the edges are sequences of rectilinear segments. An *orthogonal drawing* is such that every segment is *horizontal* or *vertical*. In an orthogonal drawing the length of an edge is the number of grid points on it. Let Γ be a straight-line grid drawing and consider the smallest rectangle with sides parallel to the x- and y-axes that covers Γ completely. We call such rectangle *bounding box*. The *area* of a drawing is the number of grid points contained in its bounding box (including the border).

Let $d(p_1, p_2)$ denote the Euclidean distance between two points p_1 and p_2 in the plane. Let P be a set of distinct points in the plane; the *minimum spanning tree* of P, $\mathrm{MST}(P)$, is a connected straight-line drawing of a tree that has P as vertex set and that minimizes the total edge length. Notice that $\mathrm{MST}(P)$ is not uniquely defined, since there can be several minimum spanning trees for the same set of points.

Given a radius $r \in \mathbb{R}^+$ (called *power of transmission* or *radius of visibility*), the set $N_r(p)$ of neighbors of a point $p \in P$ is the set of points $p' \in P$ such that $d(p, p') \leq r$ and $p \neq p'$. The UDG (*unit disk graph*) of P is the graph that has a vertex for each node in P and has an edge between p_1 and p_2 if $p_2 \in N_r(p_1)$, with $p_1, p_2 \in P$. The *unit disk drawing* $\mathrm{UDD}_r(P)$ is such that there is a segment between two points p_1 and p_2 that belong to P if $d(p_1, p_2) \leq r$ (see Fig. 1 (a–b)).

The *local minimum spanning tree minus (plus) of P* is defined as follows (see Fig. 1 (c–f)):

$$\mathrm{LMST}_r^-(P) = \{(p_1, p_2) \in P \times P : (p_1, p_2) \in \mathrm{MST}(N_r(p_1)) \cap \mathrm{MST}(N_r(p_2))\}$$

$$\mathrm{LMST}_r^+(P) = \{(p_1, p_2) \in P \times P : (p_1, p_2) \in \mathrm{MST}(N_r(p_1)) \cup \mathrm{MST}(N_r(p_2))\}$$

Notice that $\mathrm{LMST}_r^-(P) \subseteq \mathrm{LMST}_r^+(P) \subseteq \mathrm{UDD}_r(P)$. A consequence of the results in [3,14] is that $\mathrm{LMST}_r^-(P)$ and $LMST_r^+(P)$ drawings are planar, have degree at most 6, do not contain any triangle, and if $\mathrm{UDD}_r(P)$ is connected, then $\mathrm{LMST}_r^-(P)$ and $LMST_r^+(P)$ are also connected. Note that, despite of their names, $\mathrm{LMST}_r^-(P)$ and $LMST_r^+(P)$ may have cycles.

A graph G is LMST⁻ (LMST⁺) *drawable* if there exists a set P of points in the plane and a value $r \in \mathbb{R}^+$ such that $\mathrm{LMST}_r^-(P)$ (resp. $\mathrm{LMST}_r^+(P)$) is a drawing of G. We call such a drawing LMST⁻ (resp. LMST⁺) drawing. Often, our results, and the motivations for these results, are the same both considering LMST⁻ or LMST⁺ drawability. In these cases we will talk about LMST drawability and about LMST drawings.

In [3,14] it has been shown that the maximum degree for a LMST drawable graph is 6, and when it is 6 the LMST is not uniquely defined. In this paper we mainly concentrate on graphs with uniquely defined LMSTs, therefore in most cases we restrict to graphs whose maximum degree is at most 5. However, the commonly adopted method to deal with cases where the LMST is not uniquely determined, is to use *tie-breaking* rules based on vertex identifiers. Hence, in the following, when we refer to cases where LMSTs are not uniquely determined, we will specify a tie-breaking rule.

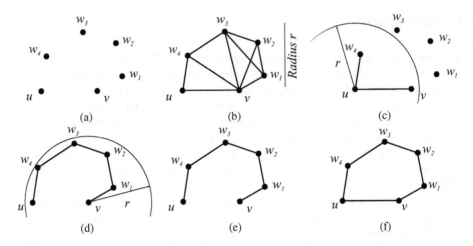

Fig. 1. (a) A set P of points. (b) $\text{UDD}_r(P)$. (c) $\text{MST}(N_r(u))$. (d) $\text{MST}(N_r(v))$. (e) LMST_r^-. (f) $\text{LMST}_r^+(P)$.

We can refine the above definitions by focusing on the role of the radius in the LMST drawings. Namely, we say that a graph is ρ-LMST *drawable* if the ratio between the length of the shortest edge of the drawing and the selected radius is ρ.

3   LMST Drawable and Non-LMST Drawable Graphs

There is a simple counterexample that shows that it is not sufficient for a graph to be planar, connected, and free of triangles to be LMST drawable.

Theorem 1. *There exists a planar, connected, and triangle-free graph with maximum degree 3 that is not LMST drawable.*

Proof. Let $G = (V, E)$ be the graph shown in Fig. 2. For the symmetry of the graph, without loss of generality, we can assume the vertices $1, 2, 4$, and 5 on the outer face (refer again to Fig. 2). Now suppose you have an $\text{LMST}_r^+(P)$ drawing or an $\text{LMST}_r^-(P)$ drawing Γ preserving such embedding of G, where P denotes the set of points where the vertices of V are mapped and r denotes the radius. Draw the segments $\overline{23}$ and $\overline{34}$.

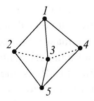

Fig. 2. A planar, connected, and triangle-free graph with maximum degree 3 that is not LMST drawable

Angles $\widehat{132}$, $\widehat{134}$, $\widehat{435}$, and $\widehat{235}$ sum to 2π, so one of them is greater than or equal to $\frac{\pi}{2}$. Let $\widehat{a3b}$ be such angle, with $a \in \{1,5\}$ and $b \in \{2,4\}$. Since in a triangle the longest side is opposite to the largest angle, then $\overline{ab} \geq \overline{a3}$ and $\overline{ab} \geq \overline{b3}$. Hence, edge (a,b) cannot be neither in $\mathrm{MST}(N_r(a))$ nor in $\mathrm{MST}(N_r(b))$ and so it cannot be in Γ, that is a contradiction. So G is not LMST drawable. $\qquad\square$

In the remainder of this section we present several classes of graphs that are LMST drawable.

Theorem 2. *All trees with maximum degree 5 are $\frac{1}{2^{n^2}}$-LMST drawable.*

Proof. From the result of Momma and Suri [19], we know that every tree T with maximum degree 5 can be drawn in the plane as the minimum spanning tree of its vertices. Suppose you have such a drawing Γ of T on a point set P in the plane. If the radius of visibility is larger than the maximum distance between two points in P, then for each vertex u we have $\mathrm{MST}(N_r(u)) = MST(P)$. This implies that $LMST_r^{+/-}(P)$ is Γ. Further, the construction in [19] is such that the ratio between the longest and the shortest edge is 2^{n^2}. $\qquad\square$

Lemma 1. *The orthogonal grid is 1-LMST drawable.*

Proof. Draw the orthogonal grid as usual (see Fig. 3 (a)). Let u be the minimum distance between two points of the grid and let H be the set of grid points. By setting the radius of visibility r to a value such that $u \leq r < u\sqrt{2}$, we obtain that every $p \in H$ has $\mathrm{MST}(N_r(p))$ that is the subgraph of the grid induced by p and its neighbors. This implies that $LMST_r^{+/-}(H)$ is the orthogonal grid. $\qquad\square$

Theorem 3. *Every graph homeomorphic to the orthogonal grid is LMST drawable.*

Proof. Suppose G is a graph homeomorphic to an orthogonal grid with n rows and m columns. First, place the vertices of the $(n \times m)$ grid as described in Lemma 1. Denote by $v_{i,j}$ the vertex at row i and column j, with $1 \leq i \leq n$ and $1 \leq j \leq m$. For each vertex $v_{i,j}$, the sum $i + j$ may be even (*even vertex*) or odd (*odd vertex*). Each edge e of the grid is incident to one even vertex and one odd vertex. When replacing e with a path, the new vertices can be inserted closer to the even vertex than to the odd vertex. It is easy to see (see Fig. 3 (b)) that the LMST of the obtained set of vertices is G. In fact, dashed segments of Fig. 3 (b) are longer than the radius r, while dotted segments are not part of the LMST since for each one of them, say e, there is an axis-parallel alternative path whose segments are shorter than e. $\qquad\square$

Lemma 2. *The hexagonal grid is 1-LMST drawable.*

Proof. Draw the hexagonal grid as usual (see Fig. 3 (c)). Let u be the distance between two points of the grid and let H be the set of grid points. By setting the radius of visibility r to a value such that $u \leq r < u\sqrt{3}$, we obtain that every $p \in H$ has $\mathrm{MST}(N_r(p))$ that is the subgraph of the grid induced by p and its neighbors. This implies that the $LMST_r^{+/-}(H)$ is the hexagonal grid. $\qquad\square$

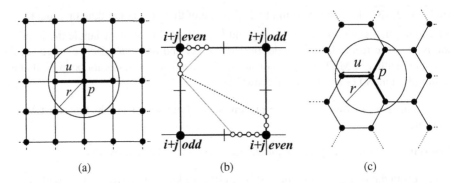

Fig. 3. LMST drawings of grids. The thick lines show $\mathrm{MST}(N_r(p))$, for a grid point p. (a) The orthogonal grid. (b) The construction used in the proof of Lemma 3. (c) The hexagonal grid.

Theorem 4. *Every graph homeomorphic to the hexagonal grid graph is* LMST *drawable.*

Proof (sketch): Let G be a graph homeomorphic to an hexagonal grid. Draw the hexagonal grid as an LMST drawing (see Lemma 2). It is easy to see that each edge can be arbitrarily split with the insertion of new vertices producing a placement whose LMST is G.

Theorem 5. *All subgraphs of the orthogonal grid are* LMST *drawable.*

For space reasons, we omit the proof and refer the interested reader to [4].

Theorem 6. *Every n-vertex outerplanar graph whose dual is a path and whose maximum degree is at most 4 is* LMST *drawable.*

Proof (sketch): Let G be a n-vertex outerplanar graph whose maximum degree is 4 and whose dual graph is a path P with s vertices. We call f_i the face of G that is dual to the i-th vertex of P. First, we remove from G all the vertices of degree two until all the faces have exactly four vertices. Let G' be the graph obtained. Draw the edge between the two vertices of degree 2 in f_1. Now we iteratively add a face that has an edge (u_1, u_2) already drawn, by drawing the vertices v_1 and v_2, where edges (u_1, v_1), (u_2, v_2), and (v_1, v_2) belong to G'. Figure 4 shows the possible cases in adding the next face, that differ one from the other depending on the degree of the vertices u_1, u_2, v_1 and v_2. Let the *growth direction* be the direction orthogonal to segment (u_1, u_2) of a face. The choice of $\epsilon = \frac{\pi}{6n}$ grants the difference between the growth direction of the first face and that of the i-th face to be bound by $\pi/2$.

We note that, by choosing parameter $\epsilon = \frac{\pi}{6n}$, the growth direction of added faces does not differ more than $\pi/2$ with respect to the growth direction of the first face. Furthermore, if we call l the length of the first edge (u_1, u_2) of f_1, it is possible to preserve the same length for edges (u_1, u_2) of all the f_i's, for $1 \leq i \leq s$, as shown in Figure 4. This gives us the possibility to choose a value for the radius equal to l to obtain that the constructed drawing is an LMST drawing. Now, we insert again the vertices

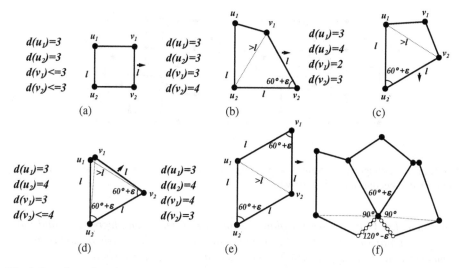

Fig. 4. Drawing a face in a outerplanar graph whose dual is a path and whose maximum degree is 4, while maintaining the LMST constraints. (a)–(e) Different cases correspond to possible degrees of the vertices u_1, u_2, v_1, and v_2. The position of the next face is indicated by an arrow. (f) The augmentation of an edge to a path in the particular case in which the edge is incident to a vertex of degree 4.

that we deleted at the beginning. For the edges incident to vertices of degree less than 4 a strategy similar to that described in the proof of Theorem 3 can be applied. Otherwise, when an edge is incident to a vertex of degree 4, Fig. 4 (f) shows how to replace the edge with a path, producing the desired LMST drawing of G.   □

4   Graph Homeomorphism and LMSTs

In this section we study the interplay between graph homeomorphism and LMST drawability.

Lemma 3. *Let G be an n-vertex planar graph whose maximum degree is 4 and that admits a planar orthogonal grid drawing with total edge length L. There exists a planar graph homeomorphic to G with at most $2L$ vertices that is 1-LMST drawable.*

Proof. Let Γ be a planar orthogonal grid drawing of G with total edge length L. We can split each horizontal and each vertical segment that is part of a polyline edge in Γ by inserting a new vertex in each grid point and half grid point (see Fig. 5). By setting the radius of visibility to a value r such that $\frac{1}{2} \leq r < 1$, it follows that the local minimum spanning tree of each vertex v consists of the subgraph induced by v and its neighbors. If we denote by P the set of points of the augmented drawing Γ, $LMST_r^{+/-}(P)$ is Γ itself. Observe that we have inserted into each edge of length l at most $2l - 3$ new vertices.   □

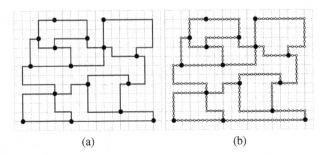

(a)   (b)

Fig. 5. (a) An orthogonal drawing of a maximum degree 4 planar graph. (b) The same drawing augmented to become an LMST drawing.

Lemma 4. *Let G be a n-vertex planar graph whose maximum degree is 4 and that admits a planar orthogonal grid drawing with area A. There exists a planar graph homeomorphic to G with at most $2A$ vertices that is 1-LMST drawable.*

Proof. Trivially, by Lemma 3 and by observing that in a planar orthogonal grid drawing with area A, the total edge length is bounded A.   ☐

Theorem 7. *Let G be an n-vertex planar (outerplanar) graph with maximum degree 4. There exists a planar graph with $O(n^2)$ ($O(n)$) vertices homeomorphic to G that is 1-LMST drawable.*

Proof. By [5] we know that every n-vertex planar graph G with maximum degree 4 admits a planar orthogonal grid drawing with $O(n^2)$ area. Further, by [7] we know that every n-vertex outerplanar graph G with maximum degree 4 admits a planar orthogonal grid drawing with $O(n)$ area. The theorem follows by applying Lemma 4.   ☐

Theorem 8. *Every n-vertex planar graph with maximum degree 5 admits a homeomorphic planar graph that is LMST drawable.*

Proof (sketch): Let G be a planar graph with maximum degree 5. Apply to G any straight-line drawing algorithm. Let d be the minimum distance between two vertices or between a vertex and an edge not incident on it. For each vertex u draw four circles c_i, with $2 \le i \le 5$, centered at that vertex, with radius $\frac{d}{12}$, $\frac{2d}{12}$, $\frac{3d}{12}$, and $\frac{4d}{12}$, respectively. Consider vertex u and suppose it has 5 incident edges (u, v_1), (u, v_2), (u, v_3), (u, v_4), and (u, v_5), in this clockwise order around u. Edge (u, v_1) remains unchanged. Draw four half-lines h_2, h_3, h_4, and h_5, starting from u with angles 72, 144, 216, and 288 degrees with respect to (u, v_1), respectively, in this clockwise order around u. Edge (u, v_i), $2 \le i \le 5$, is cut into two pieces. The part outside c_5 remains unchanged. The part inside c_5 reaches one of the circles in $\{c_2, \ldots, c_5\}$, it is then replaced by a polygonal line with vertices lying on that circle, ending with a segment straight to u. Let $\bar{d}$ be the minimum distance between the intersections between c_2 and the edges incident on u. Let $d_{min} = \min\{d, \bar{d}\}/2$. The segments of the polygonal lines have a length of at most d_{min}. Also, replace each segment that is longer than d_{min} by a polygonal line whose vertices lye on the segment and whose length is at most d_{min}. The resulting drawing is an LMST drawing with radius d_{min}.   ☐

Theorem 9. *There exists an n-vertex planar graph that does not have any homeomorphic planar* LMST *drawable graph with less than* $\Omega(n)$ *extra vertices.*

Proof. It is possible to construct such a graph by joining $n/5$ times the 5-vertex graph of Theorem 1. Each of the 5-vertex graphs requires at least 1 extra vertex to be LMST drawable, hence the statement follows.   □

5   LMST$^+$ Recognition Is NP-Hard

In this section we show that the LMST$^+$ recognition problem is NP-hard. Formally, the problem can be stated as follows:

Problem. LMST$^+$ RECOGNITION
Instance: *A graph $G(V, E)$.*
Question: *Is G* LMST$^+$ *drawable, i.e., is there a set P of points in the plane and a value $r \in \mathbb{R}+$ such that* LMST$^+$$(P)$ *is a drawing of G?*

Notice that, due to tie-breaks, given a set of points and a radius r, the corresponding LMST$^+$ may not be uniquely defined. Therefore, the definition of the LMST$^+$ RECOGNITION problem is ambiguous if a tie-breaking rule is not provided. In the following we assume that it is chosen an arbitrary (but deterministic) *tie-breaking rule*, i.e., a function $\sigma : (p_1, p_2) \rightarrow \mathbb{N}$ that assigns a different priority to each edge. When computing the MST$(N_r(v_i))$ for some vertex v_i, if more than one spanning tree with minimum edge length is found, the one that uses edges with lower priority is chosen. In the following we assume that a simple tie-breaking rule, naturally obtained from the vertex indexes, is adopted. Namely, the ordering on the vertices induces an ordering on the edges, which can be ordered first with respect to the index of their lower end-vertex, and second with respect to the higher one.

Whatever tie-breaking rule is adopted, the following lemma holds.

Lemma 5. *Let V be a set of vertices on the plane, σ be an arbitrary tie-breaking rule, and G be the* LMST$^+$ *of V with radius r. G can not have a cycle of four vertices which are at distance less or equal than r from each other.*

Proof. Suppose for contradiction that G contains a cycle of four vertices v_1, v_2, v_3, and v_4 at distance less or equal than r from each other. It follows that the four vertices are in the neighborhood N_r of each one of them. Denote by e_i, $i = 1, \ldots, 4$, the edge connecting v_i with the next vertex in the cycle. Now, pick the edge between e_1, e_2, e_3, and e_4 that is longest and has higher index according to σ. We claim that this edge can not belong to any MST$(N_r(v_i))$, $i = 1, \ldots, 4$ and therefore can not belong to G, contradicting the hypothesis. In fact, suppose without loss of generality that this longer and higher-index edge is e_1, connecting v_1 to v_2, and suppose that e_1 belongs to MST$(N_r(v_j))$ for some vertex v_j. We produce a minimum spanning tree of $N_r(v_j)$ that replaces e_1 with an edge which is shorter or has lower index. The removal of e_1 from the MST$(N_r(v_j))$ splits the tree into two connected components, one containing v_1 and the other containing v_2. Consider the path $p = e_2, e_3, e_4$. Path p starts in the component containing v_2 and ends in the component containing v_1. Hence, one between e_2, e_3, and e_4 has one end-vertex in the first component and the other end-vertex in the second component and is shorter (or has lower index) than e_1. Therefore, replacing e_1 with such an edge would produce a tree which is preferable to the $\mathrm{MST}(N_r(v_j))$, contradicting the hypothesis that e_1 belongs to it.   □

We show that LMST$^+$ RECOGNITION is NP-hard by reducing the 3SAT-3 problem to it. The 3SAT-3 problem is an NP-complete problem [8] defined as follows:

Problem. 3SAT-3
Instance: *A set C of clauses $C = \{c_1, c_2, \ldots, c_n\}$, each containing three literals from a set $X = \{x_1, x_2, \ldots, x_m\}$ of Boolean variables, such that each variable is restricted to appear at most three times.*
Question: *Is there a satisfying truth assignment for C, i.e., is there a function $V : X \rightarrow \{true, false\}$ such that the whole set of clauses evaluates to true?*

The 3SAT-3 problem is used in [2] to show that the recognition of UDGs is NP-hard, i.e., given a graph G it is NP-hard to find a placement $f : V \rightarrow \mathbb{R}^2$ such that (v_i, v_j) is in E if and only if the geometrical distance between v_i and v_j is less or equal than one. Also, based on this reduction, it is shown that even a less restrictive problem, the recognition of d-quasi UDGs, is NP-hard if $d \geq \sqrt{2/3} + \epsilon$ [11]. A *d-quasi* UDG is parameterized by $0 \leq d \leq r$ and here, E contains at least all edges between vertices that are in distance d or less, and it contains no edges between vertices that have a distance of more than r. Pairs of vertices with a distance greater than d and smaller or equal to r can be in E but they do not have to be included.

In the following we modify the reduction of [11] to show that there is a polynomial transformation from a 3SAT-3 instance to a graph that is LMST$^+$ drawable if and only if the 3SAT-3 instance is satisfiable. This proves the following:

Theorem 10. *Given a graph G it is NP-hard to decide whether it is* LMST$^+$ *drawable for some radius r.*

5.1   Graphs Representing 3SAT-3 Instances

As said above, the construction of the LMST$^+$ recognition instance starting from the 3SAT-3 instance is similar to the one used in [11]. Here, we briefly recall the steps of the construction. For a detailed description refer to [11].

First, starting from an instance C of the 3SAT-3 problem, an intermediary graph $G_C = (V_C, E_C)$ is built. A vertex in V_C represents a positive literal, a negative literal, or a clause of C. There is an edge in E_C between a vertex representing a literal and a vertex representing a clause if the literal is contained in the clause. An example for this bipartite graph representation is shown on the left of Fig. 6.

Graph G_C can be drawn on a grid of size $O(|C| \cdot |U|)$ in such a way that the literals appear consecutively on the horizontal axis and the clauses appear consecutively on the vertical axis, as shown on the right of Fig. 6.

In the grid drawing of G_C there are three different components, representing literals, clauses, and *cross-overs* of the edges. We call the connections between these components *wires*.

The grid drawing of G_C is *orientable* if its wires can be directed in such a way that: (i) for every variable at least one of its two literals has all incident wires directed away from it; and (ii) all other components (clauses and cross-overs) have at least one wire directed away from them.

In [2] it is shown that C is satisfiable if and only if the grid drawing of G_C is orientable. For example, the grid drawing of Fig. 6 is orientable. Intuitively, the first condition on the existence of a literal with all wires directed away from it guarantees that that literal can be set to false, producing a coherent truth assignment, while the second condition guarantees that each clause has a true literal and that the constraints are correctly transmitted through the cross-overs.

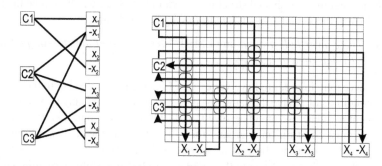

Fig. 6. On the left, G_C is shown for the instance $(x_1 \vee \overline{x}_2) \wedge (\overline{x}_1 \vee x_3 \vee \overline{x}_4) \wedge (\overline{x}_1 \vee \overline{x}_3 \vee x_4)$. On the right, a correctly oriented grid drawing is given for the same instance, representing the satisfying truth assignment $x_1 = true$, $x_2 = false$, $x_3 = false$, $x_4 = false$.

The remaining part of the reduction described in [2] consists of the construction of a d-quasi UDG that admits a realization if and only if the grid drawing of G_C is orientable. The main ingredient of this construction is a cycle, called *cage*, in which a subgraph, called *bead*, can be embedded. In [2] the beads are just chains of vertices that are attached to the cycle by a triangle, the so-called *hinge* (Fig. 7). The beads come in two sizes: a 2-bead and a 1-bead, where the first is approximately double the size of the latter. The size of cage and beads are adjusted in such a way that only one 2-bead or at most two 1-beads can be embedded within one cage.

By combining cages with each other as shown in Fig. 7 the embedding of one bead decides over the embedding of other beads in the same chain. Therefore, these chains may be used to represent directed wires. All other components, i.e., cross-overs, clauses, and literals are built out of this main building blocks, i.e., cages attached one to the other. The border of a cage can be decomposed into a sequence of eight paths, which we call N, NE, E, SE, S, SW, W, NW border-path, with obvious meaning. A cage may be attached to another cage by sharing with it one of its border-paths. Two consecutive border-paths of the same cage cannot be simultaneously used to attach to other cages.

Observe that a naive adoption of the construction used in [11] for the d-quasi UDGs would not produce a correct construction for the LMST$^+$ graph. In fact, a chain of arbitrary length is LMST$^+$ realizable in an arbitrary small area. In order to construct a

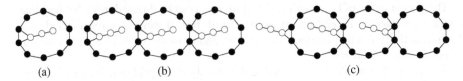

Fig. 7. (a) A cage of size 8 (black vertices) with a bead consisting of three vertices (white vertices). (b) A chain of cages in which each bead is embedded to the right of the hinge. (c) If the right bead is embedded in the middle cage it enforces the other beads to be embedded to the left.

bead gadget that has the property of not admitting an embedding in an arbitrary small area, consider first the star-shaped structure shown in Fig. 8 (a), which will be the basic building block of the 1- and 2-bead. The following lemma holds:

Lemma 6. *The convex hull around any* LMST$^+$ *realization of the star-shaped structure of Fig. 8 (a) has area at least* $\frac{6}{\sqrt{3}}r^2$.

Proof. Observe that, in order to have vertex v of Fig. 8 (a) of degree six in the LMST$^+$ graph, all neighbors of v have the same distance $L \leq r$ from v and any two consecutive edges incident to v enclose an angle of $60°$ [14]. On the other hand, vertices a and v of Fig. 8 (a) must be at distance greater than r, where r is the radius, otherwise by Lemma 5 a and v could not be part of a cycle of length four. Thus, $L > \frac{1}{\sqrt{3}} \cdot r$. It follows that the minimum convex hull around any realization of this structure has area at least $\frac{6}{\sqrt{3}}r^2 = 3.464r^2$. Also, observe that Fig. 8 (a) provides an LMST$^+$ realization for the given structure, provided that the vertices of degree 6 and 4 have lower index with respect to vertices of degree 3 and 2. □

The darker shade around the star-shaped structure in Fig. 8 (a) shows the *forbidden zone* for the structure, that is, the area in which no other vertex can be placed, unless it attaches to the structure.

Fig. 8 (b–c) shows the gadgets for the 1- and 2-beads that fulfill the needed constraints, as we state in the following lemma:

Lemma 7. *The 1-bead and the 2-bead structures shown in Fig. 8 (b–c) are such that only one 2-bead, one 1-bead, or two 1-beads can be embedded within one cage in any* LMST$^+$ *graph.*

Proof (sketch): Each cage is a cycle of the same number of vertices. The maximum area enclosed in a cage can be computed by assuming that two subsequent vertices must be at most at distance r and that the polygon they form is regular. The strategy we use to build the 1-bead and 2-bead structures is that of melting together several star-shaped structures represented in Fig. 8 (a) and that we proved in Lemma 6 to require at least a certain area in any realization. To obtain the 1-bead we melt together enough star-shaped structures until two 1-beads can be contained into a cage, but three cannot (see Fig. 8 (b) for an example).

The 2-bead shown in Fig. 8 (c) is a combination of two 1-beads joined together. Thus, the minimal area requirement of the 2-bead is the same as two 1-beads and, therefore, no other bead can share the same cage with a 2-bead. □

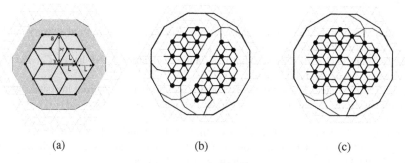

Fig. 8. (a) The star-shaped structure used in Lemma 6. The lighter shade shows the space used directly by the graph, the darker shade shows the area in which no other vertex may be placed in any realization. (b) A realization for two 1-beads embedded into the same cage. (c) A realization for a 2-bead inside a cage. Bigger vertices have lower index.

By Lemma 7, the structures of Fig. 8 can be used to replace the cages in the constructions described in [11]. Since all constructions can be performed in polynomial time, the statement of Theorem 10 follows.

6   Conclusions and Open Problems

In this paper we have shown several classes of graphs that can be represented as LMSTs, graphs that cannot, and the NP-hardness of the general problem. The following questions remain open.

- We have shown that every planar graph has a homeomorphic graph that can be drawn as an LMST. We have also given bounds on the number of vertices that should be added to the original graph to obtain drawability. Are those bounds (at least asymptotically) tight?
- We have introduced the ρ-LMST drawability and we have argued that any tree with maximum degree 5 has an LMST representation provided that such a ratio is $\frac{1}{2^{n^2}}$. Can we do better? The same problem makes sense for many other classes of graphs.
- We have proved that LMST$^+$ recognition is NP-hard. It naturally raises the question of whether also the $LMST^-$ is NP-hard.

Acknowledgments

This article has its origins in the *BICI - First Workshop on Graph Drawing*, in Bertinoro, Italy, 2006. We thank Walter Didimo, Michael Kaufmann, and Giuseppe Liotta for organizing this workshop and posing the open problems. We also thank Franz Brandenburg and Sue Whitesides for stimulating discussions and observations on the *LMST* drawability problem.

References

1. P. Bose, P. Morin, I. Stojmenovic, and J. Urrutia. Routing with guaranteed delivery in ad hoc wireless networks. *Wireless Networks*, 7(6):609–616, 2001.
2. H. Breu and D.G. Kirkpatrick. Unit disk graph recognition is NP-hard. *Computational Geometry. Theory and Applications*, 9(1-2):3—24, 1998.
3. J. Cartigny, F. Ingelrest, D. Simplot-Ryl, and I. Stojmenovic. Localized lmst and rng based minimum-energy broadcast protocols in ad hoc networks. *Ad Hoc Networks*, 3(1):1–16, 2005.
4. P.F. Cortese, G. Di Battista, F. Frati, L. Grilli, K.A. Lehmann, G. Liotta, M. Patrignani, I.G. Tollis, and F. Trotta. On the topologies of local minimum spanning tree. Technical Report RT-001-06, Dip. Ing. Elettr. e dell'Informaz., Univ. Perugia, 2006.
5. G. Di Battista, P. Eades, R. Tamassia, and I. G. Tollis. *Graph Drawing*. Prentice Hall, Upper Saddle River, NJ, 1999.
6. G. Di Battista, G. Liotta, and S. Whitesides. The strength of weak proximity. In *Graph Drawing*, pages 178–189, 1995.
7. D. Dolev and H. Trickey. On linear area embedding of planar graphs. Technical Report Report No. STAN-CS-81-876, Department of Computer Science, Stanford University, 1981.
8. M.R. Gary and D.S. Johnson. *Computers and Intractability - A Guide to the Theory of NP-Completeness*. W. H. Freeman and Company, New York, 1979.
9. J. W. Jaromczyk and G. T. Toussaint. Relative neighborhood graphs and their relatives. *Proc. IEEE*, 80(9):1502–1517, September 1992.
10. C. E. Jones, K. M. Sivalingam, P. Agrawal, and J. C. Chen. A survey of energy efficient network protocols for wireless networks. *Wireless Networks*, 7(4):343–358, 2001.
11. F. Kuhn, T. Moscibroda, and R. Wattenhofer. Unit disk graph approximation. In *Workshop on Discrete Algorithms and Methods for Mobile Computing and Communications*, 2004.
12. F. Kuhn, R. Wattenhofer, and A. Zollinger. A worst-case optimal and average-case efficient geometric ad-hoc routing. In *ACM MobiHoc*, 2003.
13. W. Lenhart and G. Liotta. The drawability problem for minimum weight triangulations. *Theoretical Computer Science*, 270:261–286, 2002.
14. N. Li, J. C. Hou, and L. Sha. Design and analysis of an mst-based topology control algorithm. In *INFOCOM*, 2003.
15. X.Y. Li. Applications of computational geometry in wireless networks. In X. Cheng, X. Huang, and D.Z. Du, editors, *Ad Hoc Wireless Networking*. Kluwer Academic Publisher, 2003.
16. X.Y. Li, G. Calinescu, P.J. Wan, and Y. Wang. Localized delaunay triangulation with application in ad hoc wireless networks. *IEEE Transactions on Parallel and Distributed Systems*, 14:1035–1047, 2003.
17. X.Y. Li, W.Z. Song, and W. Wang. A unified energy efficient topology for unicast and broadcast. In *Proc. MobiCom'05*, 2005.
18. X.Y. Li, I. Stojmenovic, and Y. Wang. Partial delaunay triangulation and degree limited localized bluetooth scatternet formation. *IEEE Transactions on Parallel and Distributed Systems*, 15(4):350–361, 2004.
19. C. L. Monma and S. Suri. Transitions in geometric minimum spanning trees. *Discrete & Computational Geometry*, 8:265–293, 1992.
20. R. Pinchasi and S. Smorodinsky. On locally delaunay geometric graphs. In *Proc. 20th ACM Symposium on Computational Geometry (SoCG'04)*, pages 378–382, 2004.
21. G. Toussaint. Geometric proximity graphs for improving nearest neighbor methods in instance-based learning and data mining. *International Journal of Comput. Geom. and Applications*, 15(2):101–150, 2005.

Distributed Routing in Tree Networks
with Few Landmarks

Ioannis Z. Emiris[1,*], Euripides Markou[1,*], and Aris Pagourtzis[2,**]

[1] Dept. Informatics & Telecoms, National University of Athens, 15784 Greece
{emiris, emarkou}@di.uoa.gr
[2] School of Elec. and Comp. Eng., National Technical University of Athens,
15780 Greece
pagour@cs.ntua.gr

Abstract. We consider the problem of finding a short path between any two nodes of a network when no global information is available, nor any oracle to help in routing. A mobile agent, situated in a starting node, has to walk to a target node traversing a path of minimum length. All information about adjacencies is distributed to certain nodes called landmarks. We wish to minimize the total memory requirements as well as keep the memory requirements per landmark to reasonable levels. We propose a landmark selection and information distribution scheme with overall memory requirement linear in the number of nodes, and constant memory consumption per non-landmark node. We prove that a navigation algorithm using this scheme attains a constant stretch factor overhead in tree topologies, compared to an optimal landmark-based routing algorithm that obeys certain restrictions. The flexibility of our approach allows for various trade-offs, such as between the number of landmarks and the size of the region assigned to each landmark.

1 Introduction

Our motivation is to model navigation in unknown or time-dependent environments, and to face questions of robustness in network routing by avoiding to store global information in a single node, or in some external oracle. Our work should find application in distributed computation and in sensor networks.

Efficient routing in a network has been studied extensively under several scenarios with respect to the kind and amount of available topology information. Several models have been proposed in order to design efficient routing schemes with relatively low memory requirements; cf. [15].

Hierarchical routing has been proposed in [20] where the problem of efficiency-memory tradeoffs for routing schemes was first raised. The authors proposed the general approach of hierarchically clustering a network into k levels and using

* Research partly supported by PYTHAGORAS, project 70/3/7392 under the EPEAEK program funded by the Greek Ministry of Educational Affairs and EU.
** Research partly supported by the General Secretariat of Research and Technology of Greece through PENED 2003 programme, contract nr. 03ED285, co-funded by the European Social Fund (75%) and national resources (25%).

T. Erlebach (Ed.): CAAN 2006, LNCS 4235, pp. 45–57, 2006.
© Springer-Verlag Berlin Heidelberg 2006

the resulting structure for routing. The total memory used is $O(n^{1+1/k} \log n)$ while bounds were derived on the increase of average path length due to the reduction of routing information. However, in order to apply the method of [20], one needs to make some fairly strong assumptions regarding the existence of a certain partition of the network. Several variations and improvements were studied later [21,25]. In [28], a landmark hierarchy has been proposed where the nodes inside the radius of some landmarks have routing information of how to send a message there. The authors give some results comparing average cases for space requirements and stretch factors. A hierarchy within a geometric setting has been proposed in [7]. In [18], the problem of minimizing the number of landmarks was studied so that the distances to the landmarks uniquely determine a mobile agent's position.

Another approach stores limited information about the network in every node. Explicit and implicit algorithms have been proposed [26,10,11,12]. In the former kind, names and labels are arbitrary and some detailed routing information for all destinations is maintained per node. In implicit solutions [13,27], names and labels are assigned according to a scheme so that the information implicit in the labelling can be used to choose the neighbor to which a message should be sent. Results about name-independent routing schemes are found in [4,23,3].

Lower bounds for the space-efficiency tradeoff of routing schemes were studied in [24,8,9,16,6,22,14]. In particular, in [24], it has been shown that no routing strategy can guarantee for every graph a routing scheme with a stretch factor $O(k)$ and $o(n^{1+1/k})$ bits of total memory. Other lower and upper bound results for space-stretch tradeoff can be found in [1,2].

A related problem is the k-center problem, where one has to select k centers and to partition the nodes among them so that no more than L nodes be assigned per center, and the distance between a node and its center be minimized. There exist constant-ratio approximation algorithms for this problem, e.g. in the works cited below. Although this literature is very rich, we do not apply their algorithms in a black-box manner and, instead, we propose our own distribution scheme. The reason is that such algorithms either do not put a limit to the region size [17], or, whenever they do, as in the capacitated k−center problem [5,19], they do not require that every region be connected; thus, a path from a node to its center can cross into another region. However, these two requirements are essential in reducing memory load. Our scheme on trees guarantees both. Another reason for preferring the term "landmark" to "center" is that the former suggests that landmarks are visible within their region and beyond. With tree networks this is true because there is a natural direction (namely, upwards) so as to reach a landmark.

Formally, given parameter L, we define certain nodes as landmarks and partition the nodes to corresponding regions so that no more than L nodes are assigned per landmark. A mobile agent MA with limited storage and computing capabilities, follows a simple algorithm using information stored at nodes it traverses in order to reach quickly a target node. One faces three simultaneous (and contradictory) tasks while distributing information about the graph:

- only a limited number of landmarks should contain information, while every other node may know only its label,
- the total memory needed should be minimized,
- the stretch factor (travelled / optimal distance) of the routing scheme should be minimized.

In this paper we propose Tree-Landmarks Routing (TLR), which is a distributed routing scheme appropriate for tree networks the nodes of which have labels assigned in a DFS manner. We store additional information (apart from node labels) only to a certain number of nodes. We also assume that every node knows the port leading to its parent. We prove the efficiency of our approach, namely that it stores a linear total amount of information to landmarks. We also show that TLR yields a constant overhead on the stretch factor, compared to that of an optimal landmark-based routing scheme that obeys the same upper and lower bounds on the size of regions. Our algorithm is flexible thus allowing for various trade-offs: most importantly between the number of landmarks and the region size. In particular we prove that, given an upper bound L on the region size and a parameter, TLR partitions a given tree network to $O(\sqrt{\frac{\delta}{\alpha}}\frac{n}{\sqrt{L}})$ regions of size at most αL, where δ is the fan-out of the tree, for any $\alpha \geq 1$. We expect that our approach generalizes to arbitrary graphs; we offer indications of this generalization.

The rest of the paper is organized as follows. The next section formalizes the problem and introduces all necessary notions. Section 3 focuses on tree networks, describes our routing scheme and proves its properties. Section 4 shows that our scheme achieves a low stretch factor and constant stretch overhead compared to routing schemes that follow similar assumptions; a second overhead measuring model is also examined. We conclude in Section 5 by suggesting possible extensions of our approach to arbitrary graphs; we also state some open questions.

2   Model and Preliminaries

Let us describe the general framework before concentrating on trees. A mobile agent MA, situated in a starting node s of a graph $G = (V, E)$, has to walk to a target node t of G traversing a minimum number of edges (or a walk of a minimum weight). The goal is to distribute information about the graph topology to a small subset of nodes so that MA can find quickly its way towards any target node.

More specifically, in a precomputation phase, among the n nodes of G, k of them are selected as *landmarks* and every node v of the graph is associated with exactly one landmark $m = lm(v)$ so that the following property is respected: The selection of landmarks is such that no landmark has more than L associated nodes. After the selection of landmarks and the association of nodes to landmarks, we distribute the information about the graph as follows.

First, in every node v of G, the next node in the path $\langle v \rightarrow lm(v) \rangle$ is stored (for the case of trees, only the port label leading to the parent of a node is necessary). Notice that the nodes have to be assigned to landmarks in such a way that, going from the node to its landmark only nodes assigned to the same landmark are encountered. Otherwise the above limited information could lead to unwanted delays, or even worse, could make MA enter a loop, thus preventing her from reaching the landmark.

Second, in every landmark m, there is a table containing information about the path $\langle m \to u \rangle$, for every node u such that $m = lm(u)$. These nodes constitute the *region* of landmark m, denoted as $region(m)$. In addition, a landmark may contain information about paths leading to some other landmarks together with indication about which nodes belong to the regions of those landmarks.

We say that two landmarks m_1, m_2 are *adjacent* or *neighbors* if there are adjacent nodes u, v so that: $m_1 = lm(u)$, $m_2 = lm(v)$.

We will mainly deal with rooted trees, so let us introduce some appropriate notation. For two nodes u, v in a rooted tree T, their *least common ancestor* $lca(u, v)$ is defined in the usual way, as the common ancestor of u and v that has the greatest depth in T. We define the *least common landmark* of u and v, denoted by $lclm(u, v)$, as the common ancestor of u and v that is a landmark and that has the greatest depth in T. It can be shown that $lclm(u, v) = lm(lca(u, v))$.

The agent initially knows only s, t. A routing from s to t could be the following: MA follows a walk from s to $lm(s)$, then from $lm(s)$ to $lm(t)$ and, finally, from $lm(t)$ to t. In the first phase MA reaches $lm(s)$ using only ports leading from nodes to their parents. In the second phase it reaches $lm(t)$ using only information stored in landmarks. In the third phase it only uses information stored in $lm(t)$ to find t. For the second-phase routing it makes sense to define a hierarchy of landmarks, or use of an implicit hierarchy, as in trees.

We assume that the labels have been assigned to nodes in a DFS manner and that each node knows its label and the port leading to its parent. We also assume that MA has limited memory and computing capabilities.

We would like to have a simple and efficient routing scheme, at the same time respecting bounds on the allowed memory load of landmarks and simple nodes. We measure the efficiency of our method by means of the stretch factor and the stretch overhead. Below we define these notions formally.

Let $w_A(s, t)$ be the walk from s to t followed by a routing scheme A and let $p(s, t)$ be the shortest path from s to t. The *stretch factor* $\mathcal{SF}_{A,G}$ of routing scheme A on graph G is the maximum ratio between $|w_A(s, t)|$ (the length of $w_A(s, t)$) and $|p(s, t)|$ taken over all pairs (s, t) of nodes of G.

We also use another natural measure of performance which we call *stretch overhead* of routing scheme A on graph G and denote it as $\mathcal{H}_{A,G}$: It is the maximum ratio between $|w_A(s, t)|$ and $|w_{A^*}(s, t)|$, taken over all pairs (s, t) and routing schemes A^* that follow similar restrictions as A. In particular, we are interested in routing schemes which also use landmarks, obey the same bounds on region size as A, and follow the same routing scenario, that is, MA travels from s to t by first traveling from s to $lm(s)$, then from $lm(s)$ to $lm(t)$, and finally from $lm(t)$ to t. Note that the stretch overhead can also be defined as the maximum ratio between the stretch factors $\mathcal{SF}_{A,G}$ and $\mathcal{SF}_{A^*,G}$, taken over all similarly restricted routing schemes A^*.

3   Distributed Routing in Arbitrary Trees

In this section we present a distributed routing scheme, which we call Tree-Landmarks Routing (TLR). This scheme consists of three algorithms: an algorithm for selecting landmarks in arbitrary trees, an algorithm for distributing routing information to the nodes, and an agent navigation algorithm.

3.1   Landmark Selection Algorithm

Given a tree T and an upper bound L on the region size, the algorithm presented below selects landmarks in T and assigns regions of T to the landmarks, in such a way that the size of each region is between $L' = \sqrt{L/\delta}$ and L, where δ is the maximum fan-out of tree T. Consequently, the number of landmarks is at most n/L', that is, at most $\sqrt{\delta L}$ times the best possible (n/L) with respect to the upper bound L.

Algorithm Landmark-Selection(tree T, region size L)
select a root r_0 for T arbitrarily
set $R := \{r_0\}$
set $L' := \lfloor \sqrt{\frac{L}{\delta}} \rfloor$ (* the choice of L' will be justified in Prop. 1; we assume $L > \delta$ *)

    (* landmark selection – assignment of regions to landmarks *)
while $R \neq \emptyset$ **do**
        pick an element r from R
        add r to the set of landmarks and remove it from R
        traverse subtree $T(r)$ rooted at r in BFS manner
        until L' nodes have been visited or there are no nodes left in $T(r)$
        assign visited nodes of $T(r)$ to $region(r)$
        add unvisited nodes that are adjacent to $region(r)$ to R
end while

    (* merging of non-full regions with parent regions *)
for each r such that $|region(r)| < L'$ **do**
        let r' be the landmark of $parent(r)$ (* $parent(r) \in region(r')$ *)
        assign all nodes in $region(r)$ to r'
        remove r from the set of landmarks
end for

The algorithm builds regions of size L' starting from the root of the tree and traversing the tree in a BFS manner. When all nodes have been visited, the algorithm assigns each region of size smaller than L' to the landmark of its parent region. We will show in Prop. 1 that the selection of L' guarantees that the size of any augmented region does not exceed L. We will also give an upper bound on the number of landmarks selected by our algorithm. We first state a simple lemma.

Lemma 1. *The region of nodes assigned to a landmark r by algorithm Landmark-Selection forms a subtree rooted at r.*

Proposition 1. *Algorithm Landmark-Selection partitions a tree into regions of size at most L. The total number of regions is $O(\sqrt{\delta}\frac{n}{\sqrt{L}})$, where δ is the maximum fan-out of tree T.*

Proof. Let the *border* $b(S)$ of a region S be the set of nodes in this region that have children outside the region. Due to the fact that regions are formed by using BFS traversal, the border of S may contain: a) leaves of S and b) at most one internal node of S, namely the parent of the leaf that was included in S last.

A region S of size L' can be augmented by small (i.e. of size $< L'$) regions rooted at nodes that are adjacent to nodes in the border of S. We call these regions *child regions* of S and their landmarks *child landmarks* of S. Let us now give an upper bound on the number of child landmarks.

Clearly, the number of child landmarks of S can be at most $|b(S)|\delta$. We distinguish between two cases:

- The root of S does not belong to $b(S)$. In this case the number of child landmarks of S is at most $(L' - 1)\delta$.
- The root of S belongs to $b(S)$ (note that this can only happen if $L' \leq \delta$). In this case S is a tree of height 1, consisting of its root and $L' - 1$ leaves. Therefore, the root may have at most $\delta - (L'-1)$ children outside S; together with the children of leaves of S we get an upper bound of $(L' - 1)\delta + (\delta - (L' - 1)) = L'(\delta - 1) + 1$ on the number of child landmarks of S.

The above give an unconditional upper bound of $L'\delta - \min(L' - 1, \delta) \leq L'\delta$ on the number of child landmarks of S.

An upper bound on the size of S' (S after augmentation) is given by

$$|S'| \leq (\# \text{ child landmarks of } S)(L' - 1) + L'$$
$$\leq L'^2\delta - L'(\delta - 1) \leq L'^2\delta \leq L$$

where the last inequality holds because $L' = \lfloor\sqrt{L/\delta}\rfloor$.

Note that $L' \geq (\sqrt{\frac{L}{\delta}})/2$ since $L > \delta$. Therefore, the number of regions (also of landmarks) is at most $\frac{n}{L'} \leq \frac{2n}{\sqrt{\frac{L}{\delta}}} = O(\sqrt{\delta}\frac{n}{\sqrt{L}})$.

Region size relaxation. We can relax the maximum size of a region by means of a parameter α; in particular, for any $\alpha > 1$ we may specify αL as the maximum region size. Then, the upper bound on the number of landmarks will become $\sqrt{\frac{\delta}{\alpha}}\frac{n}{\sqrt{L}}$, that is, it will be reduced by $\sqrt{\alpha}$. Notice also that by setting $\alpha = L\delta$ we can guarantee an optimal (with respect to an 'ideal' region size L) number of landmarks n/L, at a cost of allowing the size of a region to increase up to δL^2.

Remark: We have assumed that $L > \delta$. For cases in which $L \leq \delta$ it is not clear what would be a reasonable landmark selection strategy. For example, consider a star with n nodes. For $L < n$, it turns out that there would be at most one region with more than one nodes (the one containing the center), and several regions consisting of a single landmark (recall that we insist in keeping regions connected). Therefore, the number of landmarks would be $\Omega(n)$, which would lead to poor memory efficiency (not better than that of any standard routing method).

3.2   Memory Requirements

This section details the routing information assignment scheme. The information distribution algorithm is as follows:

Starting from the root, traverse the tree in a DFS manner and assign labels to nodes. Also, assign port labels to edges: for every node u assign port labels $1, \ldots, d(u)$ to its outgoing edges, where $d(u)$ is the degree of u. In particular, assign port label 1 to the edge leading from u to its parent.

Store the information of the graph topology as follows:

- **Type 1 info (all nodes):** At every node u only its label has been stored; port 1 leads to $parent(u)$. Note that if u is not a landmark this port is the one that leads to the next node in the path $\langle u \to lm(u) \rangle$; if u is a landmark, this port leads to the next node in the path to its parent landmark.
- **Type 2 info (landmarks only):** At every landmark m, for each node v in the region of m store the label of v (for simplicity we will call it also v), the label of v's parent, and the port j which leads from $parent(v)$ to v. More precisely, the entry for v is $[v : parent(v), j]$; if v is a child of m then the entry is $[v : m, j]$.
- **Type 3 info (landmarks only):** At every landmark m, for any child landmark m' of m store $[m' : parent(m'), j \mid R_{m'}]$, where j is the port leading from $parent(m')$ to m', and $R_{m'}$ is the node with the greatest label among all nodes in the subtree rooted at m'.

In the next section we will describe how the above information can be used by an agent in order to efficiently find its way from any node s to any node t. We now show the space requirements of the above scheme in a real RAM model, where labels consume $O(1)$ space.

Proposition 2. *The above routing table assignment scheme requires $O(\delta L)$ space for each landmark. The entire space needed is $O(n)$.*

Proof. The space needed at a landmark is as follows:
- $O(1)$ for type 1 info, as above.
- $O(L)$ for type 2 info (since there are at most L nodes associated with a landmark and for each one a constant number of labels needs to be stored),
- $O(\delta L)$ for type 3 info (since there are at most δL child landmarks of a landmark m and for each of them a constant number of labels needs to be stored).

Finally, the total space needed for the information stored in the tree is $O(n)$ because:

- for type 1 info we need to store $O(n)$ port numbers in total,
- for type 2 info we need to store $O(|region(m)|)$ labels at each landmark m, which gives a total of $O(n)$ labels
- for type 3 info we need to store $O(\#$ child landmarks of $m)$ labels at each landmark m, which gives a total of $O(\#$ landmarks$) = O(\sqrt{\frac{\delta}{L}}n)$ labels, by Proposition 1.

3.3   Navigation Algorithm

We now describe a navigation algorithm that MA can use to find its way from node s to node t. In the following we denote by u the current node where MA is located (initially $u = s$); we describe the appropriate action of MA, depending on the kind of u and the routing information obtained so far. We use a boolean variable *direction* which takes values 'up' or 'down', showing MA should move upwards or downwards. Initially *direction* is set to 'up'.

- **Case a.** If u is not a landmark and direction is 'up', then go to $parent(u)$.
- **Case b.** If u is not a landmark and direction is 'down', then follow the first port in the list of ports obtained in previous steps (see below), and remove this port from the list. (In this case MA moves towards destination t — if not already there — or towards a landmark m' containing t in its subtree.)
- **Case c.** If u is a landmark, look at type 2 info table in order to find t. If t is found, set *direction* to 'down', and use table information to construct a sequence $ports(u \to t)$ of ports leading from u to t. Store $ports(u \to t)$ to local (agent's) memory, and follow the first port in $ports(u \to t)$ (removing this port from the sequence).
- **Case d.** If u is a landmark but t is not found within type 2 info table, look at type 3 info table for entries $[m' : v, j \mid R_{m'}]$ such that $m' \le t \le R_{m'}$. If such an entry is found, then set *direction* to 'down', and use type 2 info table to construct a sequence $ports(u \to v)$ of ports leading from u to v; append port j to $ports(u \to v)$. Store $ports(u \to v) \mid j$ to local (agent's) memory, and follow $ports(u \to v) \mid j$ to eventually reach m'.
- **Case e.** If u is a landmark but t is not found in tables of type 2 or type 3 info then move towards the parent of u.

Below we show the correctness of TLR, in particular, that a mobile agent situated at node s will manage to reach node t following this scheme.

Proposition 3. *Given a graph G, a target node t and an agent MA situated at some node s of G, if G is preprocessed by using landmark selection and information distribution algorithms of TLR then MA will eventually reach t by using the Navigation Algorithm of TLR.*

Proof. We will show that the agent follows 4 particular walks in a row: $\langle s \to lm(s)\rangle$, $\langle lm(s) \to lclm(s,t)\rangle$, $\langle lclm(s,t) \to lm(t)\rangle$, and $\langle lm(t) \to t\rangle$. Note that some of these may be of zero length.

If s is a landmark then MA is already in $lm(s)$, otherwise it moves up (case a) until it reaches $lm(s)$.

If t is 'visible' from $lm(s)$ (that is, belongs to the subtree rooted at $lm(s)$) then MA is already in $lclm(s,t)$, otherwise it moves upwards (case e and then case a repeatedly, possibly repeating this pattern several times) until it reaches a landmark m such that t belongs to the subtree of m (at least one such landmark exists: the root). Clearly, $m = lclm(s,t)$.

If $t \in region(m)$ then MA is already in $lm(t)$. Otherwise, it starts moving down (case d and then case b repeatedly) to reach the child landmark of m that contains t in its subtree. This process is possibly repeated several times until MA reaches $lm(t)$.

It then follows the path implied by the routing information of $lm(t)$ (case c and then case b repeatedly) in order to reach t.

All the calculations needed by MA in order to determine each time the port by which it has to leave its current position can be done in time linear in the number of nodes.

4   Efficiency of TLR Scheme

In this section we give estimations for the efficiency of Tree-Landmarks Routing (TLR) in terms of the stretch factor and the stretch overhead achieved.

4.1   Stretch Factor Analysis

In the following we establish a stretch factor for TLR in trees which is proportional to the region height.

Proposition 4. *TLR achieves stretch factor $\mathcal{SF}_{TLR,T} = 2h + 1$ on a tree T, where h is the maximum height of a region of T.*

Proof. Consider a pair of nodes s, t. The walk followed by MA under TLR scheme can be analyzed in the following paths (in this order): $\langle s \to lca(s,t) \rangle$, $\langle lca(s,t) \to lclm(s,t) \rangle$, $\langle lclm(s,t) \to lca(s,t) \rangle$, and $\langle lca(s,t) \to t \rangle$. Clearly, all these paths are shortest paths since they are simple. Therefore,

$$\mathcal{SF}_{TLR,T} = \max_{s,t} \frac{|p(s, lca(s,t))| + 2|p(lca(s,t), lclm(s,t))| + |p(lca(s,t), t)|}{|p(s,t)|}$$

$$\leq 1 + \max_{s,t} \frac{2|p(lca(s,t), lclm(s,t))|}{|p(s,t)|}$$

because the shortest path $p(s,t)$ between s and t consists of shortest paths $p(s, lca(s,t))$ and $p(lca(s,t), t)$.

The distance $|p(lca(s,t), lclm(s,t))|$ is at most h, since $lclm(s,t)$ is the landmark of $lca(s,t)$. Therefore, $\mathcal{SF}_{TLR,T} \leq 2h + 1$. On the other hand, there is a case in which $|p(lca(s,t), lclm(s,t))| = 2h$ and $|p(s,t)| = 1$. Namely, whenever s is parent of t (therefore $s = lca(s,t)$), and s and t are in different regions (t is a child landmark of $lm(s)$).

Taking into account that h is bounded by the maximum region size we obtain the following.

Corollary 1. *Tree-Landmarks Routing achieves a stretch factor of $O(L)$.*

4.2 Stretch Overhead with Upper and Lower Bounds on Region Size

Next we establish a constant stretch overhead for TLR in trees where all regions have size between L' and L. This restriction is necessary in order to have a fair comparison to an optimal routing scheme. Notice that, for any routing scheme under consideration, since each region size is between L' and L, the number of landmarks is between n/L and n/L'. We first prove the result for line graphs and full δ-ary trees and then sketch a proof for arbitrary trees.

Proposition 5. *TLR achieves stretch overhead 2 when applied to line graphs.*

Proof. TLR, when applied to a line, produces regions which are also lines. All these regions have size L', except possibly for one or two 'leaf' regions which are merged with their 'parent' regions. Therefore the maximum height of a region is at most $2L'$.

As mentioned above, any routing scheme A^* that we would like to compare to TLR will produce regions of size $\geq L'$. Clearly, the stretch factor for one of these regions alone will be $\geq L'$ (consider two neighbor nodes, as far from the landmark as possible). Therefore, the stretch overhead is at most 2.

We will now show that this is also the case for full (i.e. complete) δ-ary trees.

Proposition 6. *TLR achieves stretch overhead 2 when applied to full δ-ary trees.*

Proof. Let $h_\delta(r)$ denote the height of a complete δ-ary tree with r nodes; note that $h_\delta(r) = \Theta(\log_\delta r)$. During its last stage, Landmark-Selection augments a region R by combining it with regions that have the following properties: their height is smaller than $h_\delta(L')$ and their landmarks are adjacent to nodes of R that are in distance $h_\delta(L')$ or $h_\delta(L') - 1$ from the landmark of R. Therefore, the maximum height of a region after augmentation is at most $2h_\delta(L')$.

As mentioned above, A^* makes regions of size $\geq L'$. Clearly, such a region cannot have height smaller than $h_\delta(L')$, which implies that the stretch overhead of TLR is at most 2.

Next, we sketch how to extend this result to arbitrary trees.

Proposition 7. *TLR achieves stretch overhead 4 when applied to an arbitrary tree.*

Proof. Let h denote the maximum height of a region R of a tree T produced by Landmark-Selection algorithm in its first phase (that is, before merging).

As argued in the proof of Prop. 6 the maximum region height at the end of Landmark-Selection will be at most $2h$. Consider any partitioning of T into connected regions of size at least L'; it can be shown that nodes of R cannot be accommodated into regions that all have height smaller than $\frac{h}{2}$. Therefore, there exists at least one region of height $\frac{h}{2}$, leading to a stretch overhead of 4.

4.3 Stretch Overhead with Upper Bounds on Region Size and Number of Landmarks

We now consider a different definition of stretch overhead. Namely, we allow comparison with any routing scheme that respects the same upper bound L on the region size and the same upper bound on the number of landmarks n/L'.

Propositions 5 and 6 can be adapted to show that TLR achieves optimal (within a factor of 2) stretch overhead in the extreme cases of line graphs and full trees. This is because the bound on the number of landmarks guarantees that there must be at least one region of size $\geq L'$. However, this does not hold for trees that are not complete, as shown by the following example.

Example of $\Theta\left(\frac{L}{\delta \log_\delta^2 L}^{1/4}\right)$ stretch overhead. Consider a graph consisting of a line segment S with L' nodes, followed by t complete δ-ary trees, $T_1, \ldots, T_t$, each with L' nodes; the root of T_1 is adjacent to a leaf of S, and the root of T_i, $2 \leq i \leq t$, is adjacent to a leaf of T_{i-1}. Algorithm Landmark-Selection will divide such a graph to $t+1$ regions, namely S and the T_i trees. Therefore, the stretch factor will be $\Theta(L')$ due to the height of S.

On the other hand, it is possible, using the same number of landmarks (i.e. $t+1$), to split S into t regions of height L'/t and construct one region containing all T_i's, provided that $tL' \leq L$. The maximum region height of the new partition is $h^* = \max(L'/t, t\log_\delta(L'))$; this is minimized for $t = \sqrt{L'/\log_\delta L'}$ giving $h^* = \sqrt{L' \log_\delta L'}$. Hence, the stretch overhead is equal to $L'/h^* = \sqrt{\frac{L'}{\log_\delta L'}} = \Theta\left(\frac{L}{\delta \log_\delta^2 L}^{1/4}\right)$.

5 Conclusions

In this work we have proposed Tree-Landmarks Routing (TLR), which is a distributed routing scheme appropriate for tree networks. Given an upper bound L on region size, TLR partitions a tree T to $O(\sqrt{\delta} \frac{n}{\sqrt{L}})$ regions of size at most L, achieving a reasonable stretch factor and an optimal (within a constant) stretch overhead, while storing linear amount of information to landmarks and constant information per node to non-landmark nodes.

The optimal stretch overhead of TLR is achieved for the model with both upper and lower bounds on the size of regions. An interesting open question is whether we can devise a routing scheme with constant or even logarithmic stretch overhead for arbitrary trees under the model with upper-bounded region size and number of landmarks.

A second question is whether we can reduce the number of regions to e.g. $O(\frac{n}{L})$ without affecting the efficiency of the scheme. A possible modification of our landmark-selection algorithm toward this direction is the following: instead of merging regions of size $< L'$ only with parent regions, we do the same with all regions of size $< L$ as long as no region size becomes $> L$, in a bottom-up manner. This makes larger regions, resulting to fewer landmarks. In practice, we expect that for large trees most regions will be of size close to L and only a few regions will have size L'. However, to quantify these claims we would need extra hypotheses on the topology of our tree. On the downside, such a modification shall increase the stretch factor.

Another important issue is whether TLR can be extended to work for arbitrary graphs, and what the efficiency will be. A possible approach might be the following: given a graph G, specify a BFS spanning tree T for G, rooted at a minimum eccentricity node; then apply TLR to tree T. Clearly, this would lead to the same trade-off, as that for trees, between the region size and the number of landmarks and the same memory requirements. On the other hand, it seems that the stretch factor and overhead can become arbitrarily large. Therefore, a very interesting question is whether there exist routing schemes with low stretch overhead for arbitrary graphs, under any of the two overhead measuring models.

Acknowledgements. We thank Leonidas Guibas for introducing us to sensor-networks problems and related questions on distributed routing, thus leading to this paper. We also thank Christodoulos Fragoudakis, Valia Mitsou, and Elias Tsigaridas for useful discussions. Finally we thank the anonymous referees for their helpful suggestions.

References

1. I. Abraham, C. Gavoille, and D. Malkhi. On space-stretch trade-offs for compact routing schemes. Research Report RR-1374-05, LaBRI, France, November 2005.
2. I. Abraham, C. Gavoille, and D. Malkhi. Compact routing for graphs excluding a fixed minor. In Proc. 19th International Symposium on Distributed Computing (DISC), LNCS 3724 pp. 442–456, September 2005.
3. I. Abraham and D. Malkhi. Name Independent Routing for Growth Bounded Networks. In Proc. 17th ACM Symp. Parall. Algorithms and Architectures (SPAA '05), pp. 49–55, 2005.
4. B. Awerbuch, A. Bar-Noy, N. Linial, and D. Peleg. Compact distributed data structures for adaptive network routing, Proc. of 21st ACM Symp. on Theory of Computing, pp. 230–240, May 1989.
5. J. Bar-Ilan, G. Kortsarz, and D. Peleg. How to allocate network centers. J. Algorithms 15 (1993), pp. 385–415.
6. H. Buhrman, J.-H. Hoepman, and P. Vitanyi. Optimal routing tables. In Proc. 15th ACM Symp. on Principles of Distributed Computing, pp. 134–142, May 1996.
7. Q. Fang, J. Gao, L. Guibas, V. de Silva, L. Zhang. GLIDER: Gradient Landmark-Based Distributed Routing for Sensor Networks. In Proc. 24th Conf. of IEEE Com. Soc. (INFOCOM'05), 2005.

8. P. Fraigniaud and C. Gavoille. Memory requirement for universal routing schemes. In Proc. 14th ACM Symp. on Principles of Distributed Computing, pp. 223–230, August 1995.

9. P. Fraigniaud and C. Gavoille. Local memory requirement of universal routing schemes. In Proc. 8th ACM Symp. on Parallel Algorithms and Architectures, pp. 183–188, June, 1996.

10. G.N. Frederickson, R. Janardan. Separator-Based Strategies for Efficient Message Routing. In Proc. 27th IEEE Symp. on Foundations of Computer Science, 1986 pp. 428–437.

11. G.N. Frederickson, R. Janardan. Designing networks with compact routing tables. Algorithmica 3: 171–190, 1988.

12. G.N. Frederickson, R. Janardan. Efficient message routing in planar networks. SIAM Journal on Computing 18: 843–857, 1989.

13. P. Fraigniaud, C. Gavoille. Routing in Trees. In Proc. 28th International Colloquium, ICALP 2001 Crete, Greece, LNCS 2076, 2001, pp. 757–772.

14. C. Gavoille and M. Gengler. Space-efficiency of routing schemes of stretch factor three. In Proc. 4th Int. Colloq. on Structural Information & Communication Complexity, pp. 162–175. Carleton Scientific, 1997.

15. C. Gavoille and D. Peleg. Compact and localized distributed data structures. Distributed Computing, Volume 16, Issue 2-3, Sep 2003, Pages 111–120.

16. C. Gavoille and S. Perennes. Memory requirement for routing in distributed networks. In Proc. 15th ACM Symp. on Principles of Distributed Computing, pp. 125–133, May 1996.

17. D. Hochbaum and D. B. Shmoys. A best possible heuristic for the k−center problem. Mathematics of Operations Research, Vol 10:180–184, 1985.

18. S. Khuller, B. Raghavachari, and A. Rosenfeld. Landmarks in Graphs. Discrete Applied Mathematics, Vol 70 (3), pp. 217–229 (1996).

19. S. Khuller and Y.J. Sussmann, The capacitated k-center problem. SIAM J Discrete Math 13 (2000), 403-418.

20. L. Kleinrock, F. Kamoun. Hierarchical routing for large networks: performance evaluation and optimization. Computer Networks 1: 155–174, 1977.

21. L. Kleinrock, F. Kamoun. Optimal clustering structures for hierarchical topological design of large computer networks. Computer Networks 10: 221–248, 1980.

22. E. Kranakis and D. Krizanc. Lower bounds for compact routing. In Proc. 13th Symp. on Theoretical Aspects of Computer Science, LNCS 1046, pp. 529–540, February 1996.

23. D. Peleg. Distance-dependent distributed directories. Information and Computation, pp. 270-298, 1993.

24. D. Peleg and E. Upfal. A tradeoff between size and efficiency for routing tables. J. ACM, 36:510–530, 1989.

25. R. Perlman. Hierarchical networks and the subnetwork partition problem. In Proc. 5th Conf. on System Sciences, 1982.

26. N. Santoro, R. Khatib. Labelling and implicit routing in networks. The Computer Journal 28: 5–8, 1985.

27. M. Thorup, U. Zwick. Compact routing schemes. In Proc. 13th annual ACM symposium on Parallel algorithms and architectures, pp. 1–10, 2001, Crete Island, Greece.

28. P.F. Tsuchiya. The landmark hierarchy: A new hierarchy for routing in very large networks. Computer Communication Review 18, 4 (August 1988), pp. 35–42.

Scheduling of a Smart Antenna: Capacitated Coloring of Unit Circular-Arc Graphs

Guy Even[1] and Shimon Shahar[2]

[1] School of Electrical Engineering, Tel-Aviv University, Tel-Aviv 69978, Israel
guy@eng.tau.ac.il
http://www.eng.tau.ac.il/ guy
[2] School of Electrical Engineering, Tel-Aviv University, Tel-Aviv 69978, Israel
moni@eng.tau.ac.il

Abstract. We consider scheduling problems that are motivated by an optimization of the transmission schedule of a smart antenna. In these problems we are given a set of messages and a conflict graph that specifies which messages cannot be transmitted concurrently. In our model the conflict graph is a unit circular-arc graph.

Two variants of the problem are considered: C-MBL and NU-C-MBL. In C-MBL, the messages have unit demands, whereas in NU-C-MBL demands are arbitrary. We present an optimal algorithm for C-MBL and a 3-approximation algorithm for NU-C-MBL.

Keywords: smart antennas, scheduling with conflicts, capacitated coloring, unit circular-arc graphs.

1 Introduction

1.1 Background: Scheduling of Smart Antennas

The major issue in wireless networks is how to utilize bandwidth efficiently. This problem is becoming more and more acute as the density and bandwidth requirements of clients increase. The only way to alleviate this bottleneck is to use resources more efficiently. For example, instead of transmitting a message to all direction in the vicinity of an antenna, use a directional beam that concentrates the energy only in the direction of the receiving client. This approach has the following advantages. First, the transmission energy is reduced, so apart from saving energy, health hazards are reduced. Second, interference is reduced as clients do not receive transmissions that are not aimed towards them. This means that reliable communication is possible with weaker signals and more messages can be transmitted simultaneously.

Traditionally, a directional beam is formed by using a reflector. To obtain a pencil shaped beam, a two-dimensional parabolic reflector (i.e. a dish) is used. The beam is directed in different directions by rotating the dish so that it is aimed in the desired direction. This mechanical method of steering the beam is slow and is appropriate for base stations that serve a small and fixed set of clients.

T. Erlebach (Ed.): CAAN 2006, LNCS 4235, pp. 58–71, 2006.

Large and mobile sets of clients are often served by organizing horn shaped antennas side by side. Each horn forms a beam that has the shape of a sector (say, a sector with an angle of 60 degrees). This enables transmission in all directions but has two main drawbacks. First, the gain due to the wide sector is small. Second, separate radio frequency (RF) processing is required for each horn antenna. This incurs an additional cost, especially since high power RF components are expensive.

Advances in technology enable forming directional beams with an array of simple antennas (instead of reflectors). The common signal is multiplied by antenna specific complex weights (i.e., each antenna in the array is associated with a different weight). The multiplication by a complex weight changes the amplitude and the phase of the signal. The signals transmitted from different antennas in the array add in interfering and accumulating patterns to form a directional beam (see left part of Fig. 1). By modifying the complex weights, it is possible to electronically steer the beam from one direction to another. Electronic steering can be performed very quickly since no mechanical movement is required. Moreover, a few directional beams in different directions can be super-positioned (see right part of Fig. 1).

Calculating the required weights needed to form a given beam is a complicated task (see: [15,13,8]). One method to overcome this problem is to pre-calculate the coefficients that yields a pencil beam of width 2Δ in any given direction θ, and use superposition to obtain a combination of pencil beams. Due to side-lobes and non-linear effects, the angular difference between super-positioned beams cannot be too small. To simplify the model, we assume that beams may not overlap, namely, the angular difference should be at least Δ. This means that two clients with an angular difference less than Δ may not be served simultaneously.

To summarize, we assume that the antenna is "smart" in the sense that in every time slot, the beam may be the superposition of at most C pencil beams, each of width 2Δ. The angular difference between two pencil beams in the same time slot must be at least Δ.

We consider a setting in which there are n clients in the vicinity of the antenna. The goal is to transmit message m_i to client i. We assume that all messages have the same length. The goal is to schedule the transmission of the messages to minimize the amount of time needed to transmit all messages. In each time slot, at most C messages can be sent simultaneously. This is achieved by pointing at most C beams towards the clients whose messages are transmitted in this time slot. Messages scheduled for transmission in the same time slot may not conflict. Namely, the angular difference between the corresponding clients must be at least Δ. We now formalize the problem of scheduling a smart antenna.

1.2   Formalizing the Model

The model in this paper is of a network with one antenna that serves a set of n clients. Both the antenna and the clients are modeled as points in the plane. The location of client i is denoted by its polar coordinate $(r(i), \phi(i))$, where the antenna is positioned in the origin. The antenna should transmit a unique

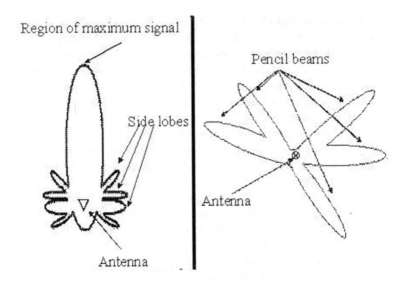

Fig. 1. (A) A pencil beam. The contour bounds a region in which the reception level is above a certain level. Note that the main lobe points up. (B) The superposition of 5 pencil beams.

message m_i for each client i. In the sequel, we do not distinguish between the client i and its message m_i, so we simply refer to $(r(i), \phi(i))$ as the location of the message m_i. The assumptions on the transmission model are the following:

- The transmission is done in time slices.
- During a time slice, the antenna transmits the messages of a subset of the clients.
- When transmitting to client i, the antenna aims a pencil beam in its direction $\phi(i)$.
- During a time slice, the antenna's beam is a superposition of multiple pencil beams. The antenna may support at most C pencil beams concurrently, provided that the angular difference between every two pencil beams is at least Δ.

  The angular difference between every two beams is at least Δ. Therefore, two clients i, j for which $|\phi(i) - \phi(j)| \leq \Delta$ are in conflict, and their messages cannot be scheduled concurrently.

Let M be the set of messages, i.e., $M = \{m_i \mid 1 \leq i \leq n\}$. Two messages i, j are *in conflict* if $|\phi(i) - \phi(j)| \leq \Delta$. A *schedule* for the antenna, is a partition of M such that: (i) $M = \bigcup_{i=1}^{k} C_i$, and (ii) every two messages in the same partition C_j are not in conflict. The *length* of the schedule, is simply the number of subsets k. The *Minimum Broadcast Length* problem (MIN-LEN), is the problem whose input is a set of messages with locations, and its output is a schedule of minimum length.

An extension to MIN-LEN problem is obtained by limiting the number of messages than can be transmitted simultaneously. This limit is caused by the specifications of the antenna's array that forms the beam. We refer to this bound as the *capacity*. The capacity C poses a constraint on a schedule, namely, $\forall j \; |C_j| \leq C$. The problem of minimizing the schedule length in the presence of a capacity constraint is called the *C-Minimum Broadcast Length* problem (C-MBL).

Another type of capacity constraint arises in a weighted scenario. Assume that different messages require different transmission power. For example, the power required for broadcasting a message m_i may depend on the distance r_i^2, or on the reception quality of the client(i.e. the client's antenna). Another motivation for this problem is dealing with different noise levels. A natural way to deal with the noise level, is to use code words of different lengths to the clients. Namely, if a client has a low noise level, then the code words that are used for transmission to this client are short; whereas if a client has a high noise level, then the code words that are used for transmission are longer. Therefore we connect the bandwidth demands of a client with his noise level. Since the antenna has a limited bandwidth, the capacity constraints is a limit on the overall bandwidth for transmission in a time slice. However, this leads to a reduction between the problem of dealing with noise level and the problem of allocating transmission power. I.e. the antenna deals with higher noise level by assigning a wider bandwidth. A client with wider bandwidth is assigned a wider range of frequencies, and hence the antenna spends more power on its message. In order to capture a general scenario, we model this problem as follows. For each message m_i, we assign a positive number d_i, that is its demand. For the antenna we assign a number C (also referred to as its capacity), that stands for its maximum power per time slice. Hence we require: $\forall j \; \sum_{i \in C_j} d_i \leq C$. We refer to the problem of minimizing the broadcast length where demands exist as the *Non-Uniform C-Minimum Broadcast Length Problem* (NU-C-MBL).

1.3   Graph Theoretic Formulation

In this section we formulate the scheduling problems presented as graph theoretic problems. We define the *collisions graph* $G = (V, E)$ as follows. The vertex set $V(G)$ contains one vertex v_i for each message m_i, the edge set $E(G)$ contains the edge (v_i, v_j) if and only if the messages m_i, m_j interfere.

Due to the geometric origin of the constraints in our problem, the collisions graph is a unit circular-arc graph. A unit circular-arc graph, is a graph that has a realization as the collision graph of a set of unit length arcs. Namely, there exists a set R of unit length arcs on a circle, and a bijection $f : V \rightarrow R$ such that: $(v_i, v_j) \in E$ if and only if $f(v_i) \cap f(v_j) \neq \emptyset$ (the arcs are intersected).

In order to avoid confusion, we would like to emphasize that an arc which represents a message, corresponds to a vertex of the collision graph. The edges of the collision graph, that in some of the literature are referred to as arcs, corresponds to intersections between arcs. Nevertheless, we preferred this choice of terms since it follows the standard terms used for circular-arc graphs.

The arc realization of vertex v_i is simply the sector of size Δ centered at $\phi(i)$, i.e., the sector $[\phi(i) - \Delta/2, \phi(i) + \Delta/2]$. Since this is a projection of the clients locations on the unit circle, two arcs $f(v_i), f(v_j)$ intersect if and only if their corresponding messages m_i, m_j interfere.

The MIN-LEN problem, can be reduced to the problem of computing a minimum coloring the graph G. Given a minimum coloring, it assigns to each arc m_i a color from the set $\{1, \ldots, \chi^*\}$, where χ^* denotes the chromatic number of the graph. This coloring defines the schedule in which the clients scheduled to time slice t are all the clients whose arcs are colored t. Due to the requirement that the messages in each time slice are independent sets, the length of this schedule is minimum.

The C-MBL and NU-C-MBL are capacitated versions of the coloring problem. E.g., in C-MBL we need to partition the $V(G)$ into minimum number of independent sets $\{C_j\}$, such that the size $|C_j|$ of each set does not exceed the antenna's capacity.

1.4   Previous Results

Two problems related to C-MBL and NU-C-MBL are coloring of circular-arc graphs and bin-packing with conflicts. We briefly review previous works on these problems.

Previous results on coloring circular-arc graphs. Circular-arc graphs are a natural generalization of intervals graphs since the removal of any maximal clique from a circular-arc graph yields an intervals graph. (This is due to the fact that the set of arcs passing for a given point on the circle forms a subset of a maximal clique.) Since intervals graphs are perfect, finding their chromatic number χ^* is a well known polynomial problem (see for example [14]). However, despite the resemblance, finding the chromatic number of circular-arc graphs is NP-Hard ([2]). The coloring problem remains NP-Hard even for special families of circular-arc graphs ([3]). Based on this connection between intervals graphs and circular-arc graphs, the following simple algorithm is a 2-approximation for coloring circular-arc graphs: (i) Color a maximal clique, (ii) remove it from the graph so as to obtain an intervals graph, and (iii) color optimally the intervals graph.

A different 2-approximation algorithm for coloring circular-arc graph was given by Tucker ([11]). His algorithm is a greedy, first-fit algorithm that has two stages: (i) first generate an order on the arcs, and (ii) assign consecutive blocks of arcs according to the same color class, according to the order generated. Tucker also conjectured that $\frac{3\omega}{2}$ colors are always sufficient to color a circular-arc graph, where ω is the size of the maximum clique. This conjecture was proved by Karapetian ([6]). Shih and Hsu gave a $\frac{5}{3}$ approximation algorithm ([10]). Their algorithm is based on extending Tucker's ideas to cases where no three arcs cover the circle. Recently Valencia-Pabon showed that if the minimum size of a dominating set is ℓ, then the approximation ratio of Tucker's algorithm is $\frac{(\ell-1)}{(\ell-2)}$ ([12]).

An approximation algorithm based on linear programming for coloring circular-arc graphs was given by Kumar ([7]). His algorithm achieves an approximation ratio of $(1 + 1/e + o(1))$ for graphs whose coloring number χ^* is $\Omega(\log n)$. His algorithm first reduces the coloring problem into a multi-commodity flow problem, then it solves the corresponding linear program, and finally it rounds up the solution.

An important family of circular-arc graphs is proper circular-arc graphs. A proper circular-arc graph is a circular-arc graph for which there exists a realization R such that no arc is fully contained within another. An even more restricted family of graphs are unit circular-arc graphs, for which there exists a realization such that all arcs lengths are the same, trivially such a realization is also proper. Orlin, Bonuccelli and Bovet ([9]) showed that coloring proper circular-arc graphs is polynomial. The running time of their algorithm is $O(n^2 \log n)$, and it is based on a procedure that checks whether a graph is k-colorable, for a given value of k. We refer to this algorithm as COL-PROPER(G).

Corollary 1. *By the reduction given in the previous section, we obtain that* COL-PROPER(G) *optimally solves the* MIN-LEN *problem.*

An important balancing property of COL-PROPER(G), is that the size of all the color classes it generates is either $\left\lceil \frac{n}{\chi^*} \right\rceil$ or $\left\lfloor \frac{n}{\chi^*} \right\rfloor$, thus the color classes have the same cardinality up to rounding.

Previous results on bin-packing with conflicts. The input to a bin-packing problem with conflicts is a set V of elements with weights between zero and one. In addition, the input contains a conflict graph $G = (V, E)$ whose vertex set is the set of elements. An edge $(u, v) \in E$ indicates that u and v cannot be assigned to the same bin. The goal is to find a packing of V into as few bins as possible, where the size of each bin is one. The NU-C-MBL problem is a bin-packing problem with conflicts where the conflict graph is unit circular-arc graph.

Jansen [5] considered bin-packing with conflicts where an optimal coloring of the conflict graph is given (e.g., coloring is polynomially solvable for a class of graphs that contains the conflict graph). For such cases, an r-approximation is presented, where $2.69 \leq r \leq 2.7$. In addition, a 2.5-approximation algorithm for graphs that can be precolored is presented[1]. The approximation algorithms are based on an analysis of the first-fit-decreasing algorithm for bin packing after small elements are reweighted and matched with big elements.

An asymptotic approximation scheme for bin-packing with conflicts was presented by Jansen [4] for d-inductive graphs[2].

Epstein & Levin [1] present improvements for bin-packing with conflicts. The improvements in [1] over [5] are due to a better choice of weights assigned to small elements and a tighter analysis. The contributions in [1] are as follows: (i) A 5/2-approximation algorithm for conflict graphs that can be optimally colored

---

[1] The precoloring extension problem is defined as follows: Given a graph $G = (V, E)$ and a subset of vertices $V' = v_1, ..., v_k$. Find a minimum coloring of of G satisfying $f(v_i) = i$ for all $i = 1, .., k$.

[2] A graph is d-inductive, if there exists an ordering of the vertices such that the "left" degree of every vertex is at most d.

in polynomial time. This implies a 5/2-approximation algorithm for NU-C-MBL. (ii) A 7/3-approximation algorithm for conflict graphs that can be optimally precolored. (iii) A 7/4-approximation algorithm for bipartite conflict graphs. (iv) An online algorithm with a 4.7-competitive ratio for conflict graphs that are interval graphs.

1.5 Our Results

In this paper we present the following results:

- An optimal algorithm for C-MBL, whose running time is $O(n^2 \log n)$.
- An $O(n \log n)$ heuristic for C-MBL based on Tucker's coloring algorithm. We bound the approximation ratio of this heuristic by 3. This heuristic is optimal under a condition that seems to hold in practice.
- NP-Hardness of NU-C-MBL.
- A very simple 3-approximation algorithm for NU-C-MBL(see [1] for better approximation).

2 The Capacitated Minimum Broadcast Length Problem

In this section we present two results. The first result is an optimal algorithm for C-MBL, to which we refer by OPT-CMBL. The second result is a heuristic, based on Tucker's algorithm, which is optimal under some assumptions that may hold in practical cases. We think this heuristic is interesting since it is simple, and its running time is $O(n \log n)$, whereas the running time of OPT-CMBL is $O(n^2 \log n)$.

2.1 An Optimal Algorithm for c-mbl

Let C denote the antenna's capacity and n the number of messages. Let G denote the collision graph (G is a unit circular-arc graph). Let χ^* be the chromatic number of G. We denote by ℓ^* the length of an optimal schedule. In this section we prove the following theorem; the proof is separated into a few claims.

Theorem 1. $\ell^* = \max\{\chi^*, \lceil \frac{n}{C} \rceil\}$. Moreover, an optimal cover can be computed in polynomial time.

We begin by proving the lower bound.

Claim. $\ell^* \geq \max\{\chi^*, \lceil \frac{n}{C} \rceil\}$.

Proof. The first bound follows since the messages transmitted concurrently in every time slice are independent. The second bound follows simply from the fact that the number of messages transmitted in any time slice is bounded by C.

The following corollary shows that if the color classes are small, then the optimal coloring induces an optimal schedule.

Corollary 2. If $\lceil \frac{n}{\chi^*} \rceil \leq C$ then COL-PROPER(G) induces an optimal schedule.

Proof. The balancing property of COL-PROPER(G) implies that the largest color class in the coloring computed by COL-PROPER(G) contains $\left\lceil \frac{n}{\chi^*} \right\rceil$ clients. Since $\left\lceil \frac{n}{\chi^*} \right\rceil \leq C$ the coloring induces a valid schedule. Since $\ell^* \geq \chi^*$, optimality follows.

We now deal with the case of large color classes.

Claim. If $\left\lceil \frac{n}{\chi^*} \right\rceil \geq C$ then $\ell^* = \left\lceil \frac{n}{C} \right\rceil$.

Proof. We first reduce the graph G so that the sizes of the color classes computed by COL-PROPER(G) are not divisible by C. The reduction proceeds as follows: (i) Partition the arcs into color classes using COL-PROPER(G). (ii) If the size of a color class C_i is divisible by C, schedule the messages in C_i in groups of size C. Denote the number of arcs assigned in this step by n'', and let $n' = n - n''$. If $n' = 0$, then we are done, and the theorem holds. Let G' denote the collision graph induced by the n' remaining nodes. Obviously, $\ell^*(G) \leq \ell^*(G') + \frac{n''}{C} = \left\lceil \frac{n}{C} \right\rceil$. It suffices to prove that $\ell^*(G') = \left\lceil \frac{n'}{C} \right\rceil$. We may therefore assume that the size of every color class computed by COL-PROPER(G) is not divisible by C. Namely, neither $\left\lceil \frac{n'}{\chi} \right\rceil$ or $\left\lfloor \frac{n'}{\chi} \right\rfloor$ is divisible by C. As a result of this reduction, we assume that the sizes of the color classes in the coloring computed by COL-PROPER(G) are not divisible by C and that $C < \left\lfloor \frac{n}{\chi^*} \right\rfloor$ (i.e., every color class contains more than C arcs).

We use the following terminology. If S is an independent set and $|S| = k \cdot C + r$. Then S can be partitioned to *C-groups* and a *leftover-group*. The C-groups are k disjoint subsets of C_i (each of size C), and the leftover group is a subset of C_i of size r.

A listing of Algorithm SCHEDULE appears as Algorithm 1. The algorithm computes an optimal balanced coloring of the collision graph G using COL-PROPER(G). The idea is to schedule a C-group from each color class. The issue of the leftover group is solved by finding a C-group in the union of the current color class and the next color class (see the Replacement Claim below) to schedule C additional arcs. Since all slots, but perhaps the last slot, contain C arcs, Algorithm SCHEDULE computes a schedule whose length is $\left\lceil \frac{n}{C} \right\rceil$. Hence, we only need to prove the Replacement Claim.

Claim (Replacement Claim). In Line 8 of Algorithm SCHEDULE there always exists a set $S \subset C_i \cup C_{i+1}$ such that: (i) $|S| = C$, and (ii) both S and $(C_i \cup C_{i+1}) \backslash S$ are independent sets.

Proof. Consider the sets C_i and C_{i+1} in Line 8. Note that the set C_i is not the original color class C_i that was generated by COL-PROPER(G), since its size is less than C. We have $0 < |C_i| < C < |C_{i+1}|$. The graph induced by C_i and C_{i+1} is 2-colorable unit circular-arc graph. Hence, the union $C_i \cup C_{i+1}$ is a cycle or a union of disjoint simple paths in G. If $C_i \cup C_{i+1}$ is a cycle, it should contain

---

Algorithm 1. Algorithm SCHEDULE - scheduling when $\left\lfloor \frac{n}{\chi^*} \right\rfloor > C$.

1: Run COL-PROPER(G).
2: **for** $i = 1$ to $\chi^* - 1$ **do**
3:     **while** $|C_i| \geq C$ **do**
4:         Schedule a C-group $T \subseteq C_i$.
5:         $C_i = C_i \setminus T$.
6:     **end while**
7:     **if** $C_i \neq \emptyset$ **then**
8:         Find an independent set $S \subset (C_i \cup C_{i+1})$ of size C such that $(C_i \cup C_{i+1}) \setminus S$ is also independent. {*Existence guaranteed by the Replacement Claim*}
9:         Schedule S
10:        $C_{i+1} = (C_{i+1} \cup C_i) \setminus S$
11:    **end if**
12: **end for**
13: Schedule the C-groups and leftover group contained in C_{χ^*}.

---

the same number of edges from C_i and from C_{i+1}, contradicting the fact that $|C_i| < C < |C_{i+1}|$. Hence, $C_i \cup C_{i+1}$ is a union of disjoint simple paths in G. In each path the difference between the number of vertices from each color classes is zero or ± 1.

We apply the following iterative "balancing" procedure to obtain S. The procedure partitions $C_i \cup C_{i+1}$ into S_1 and S_2. Initially, $S_1 = C_i$ and $S_2 = C_{i+1}$, so $|S_1| < C < |S_2|$. In each iteration, arcs are moved between S_1 and S_2 so that $|S_1|$ increases by one and $|S_2|$ decreases by one. The invariant that S_1 and S_2 are a partition of $C_i \cup C_{i+1}$ into independent sets is always kept. The balancing procedure stops as soon as the size of one of the sets S_1 or S_2 is C.

Consider an iteration of the procedure in which $|S_1| < |S_2|$. The moving of arcs proceeds as follows. Pick an odd path P in which $|S_2 \cap P| > |S_1 \cap P|$. We update the sets S_1, S_2 as follows: $S_1 = (S_1 \setminus P) \cup (P \cap S_2)$, $S_2 = (S_2 \setminus P) \cup (P \cap S_1)$.

We remark that the proof of the above claim implies an efficient algorithm to find the required set S, hence Algorithm SCHEDULE is polynomial.

In the appendix we present an efficient scheduling algorithm for the case that $\left\lfloor \frac{n}{\chi^*} \right\rfloor > 2C$. This algorithm captures the intuition that if the color classes are large, then a leftover can be augmented by arcs from the next color class to fill a time slot.

2.2   A Practical Heuristic: Spiral Scheduling

The most time consuming part of the algorithm for C-MBL is the coloring of the graph by procedure COL-PROPER(G). In this section we present an algorithm called SPIRAL that avoids the coloring step. SPIRAL is based on Tucker's algorithm for coloring circular-arc graphs. Thanks to the unit length of the arcs, the implementation of SPIRAL is easier than the implementation of Tucker's algorithm.

SPIRAL has three major steps: ordering the arcs, coloring, and scheduling. We begin by describing the ordering step. Consider a representation of a circular-arc graph using a circle and arcs of the circle. We define the *load* of a point P on the circle to be the number of arcs that contain P. We order the arcs in spiral ordering as follows: (i) Set A_1 to be an arc whose starting point has the biggest load. (ii) Initially, all arcs but A_1 are unmarked. For every i between 1 and $n-1$, set A_{i+1} to be the unmarked arc whose starting point is the closest to the end of A_i. We refer to an order generated by the sorting as a spiral ordering.

We now describe the coloring step. Given a spiral ordering $A_1, \ldots, A_n$ of the arcs, the arcs are colored by iteratively "peeling" a maximal prefix of arcs that is an independent set. Each such maximal prefix is assigned a new color. In other words, initially $\mathcal{A}$ is the set of all arcs. In iteration i, pick a maximal prefix P_i of $\mathcal{A}$ that is independent. Color P_i with the color i, and update $\mathcal{A} \leftarrow \mathcal{A} \setminus P_i$. This procedure is repeated until $\mathcal{A}$ is empty.

In the scheduling step each color class is simply partitioned into C-groups and perhaps a leftover group.

In the case of unbounded capacity, this algorithm simply reduces to Tucker's algorithm. Therefore, even for the unbounded capacity case the algorithm is not optimal. We prove that SPIRAL is a constant ratio approximation algorithm.

Claim. The approximation ratio of SPIRAL is no bigger than 3.

Proof. Let $\chi_T(G)$ denote the number of colors used in the coloring step. Let $\ell_s(G)$ denote the length of the schedule computed SPIRAL. We claim that

$$\ell_s(G) \leq \frac{n}{C} + \chi_T(G).$$

Indeed, there are at most $\frac{n}{C}$ slots with C arcs and there is at most one leftover group from each color class.

The claim now follows from Theorem 1 and from the fact that $\chi_T(G) \leq 2 \cdot \chi^*$ (this was proved by Tucker [11]).

Suppose that subsets of C consecutive arcs in the spiral ordering are independent. In this case, after spiral ordering we could peel off a prefix of C arcs in each step to obtain an optimal schedule. Hence, a variation of SPIRAL leads to an optimal schedule if this condition holds.

Consider the following two conditions: (i) $C \cdot \Delta < \pi/2$ and (ii) The number of clients is proportional to the angle of a section, provided that the angle is large (say, $\pi/4$). Under such conditions most C consecutive arcs are independent, and the actual approximation ratio of the algorithm is much better than the theoretical bound. We believe that these two conditions hold in practice.

The running time of SPIRAL is dominated by the time to compute the spiral ordering which is $\theta(n \log n)$. This is more efficient than the optimal algorithm whose running time is $\theta(n^2 \log n)$.

3   The Non-uniform Capacitated Minimum Broadcast Length Problem

In this section we deal with non-uniform demands, i.e., the NU-C-MBL problem. We prove that NU-C-MBL is NP-Hard, and present a simple 3-approximation algorithm for it.

Claim. The NU-C-MBL problem is NP-Hard.

Proof. We reduce bin-packing to NU-C-MBL. Consider a set $\{1, \ldots, n\}$ of elements, where the size of element i is $s_i \in [0, 1]$. Reduce this bin-packing instance to an NU-C-MBL instance by constructing a set of n pairwise disjoint arcs, where the demand d_i of the ith arc is s_i. Finally, set the antenna's capacity to be $C = 1$. Since there are no coloring constraints on this graph, every packing induces a valid schedule, and vice-versa.

We present a 3-approximation algorithm for NU-C-MBL called NCMBL-APP. The algorithm NCMBL-APP starts with an optimal coloring calculated by COL-PROPER(G). Next the algorithm schedules the messages in each color class C_i independently as follows: (i) Sort the messages in C_i according to their demands in non-increasing order. (ii) Partition the C_i into time slots by "peeling" off maximal prefixes with demand at most C. The algorithm could try to add messages to fill gaps in previous time slices, but the analysis does not take this into account.

The correctness of NCMBL-APP follows from the fact that each color class is an independent set. The following claim bounds the approximation ratio of the NCMBL-APP.

Claim. The approximation ratio of NCMBL-APP is at most 3.

Proof. Consider the schedule computed by NCMBL-APP. Note that every time slot, except perhaps for the last time slot in each color class, is at least half full. Indeed, let j be a time slice that is not the last time slice of the color class C_i. Let D_j be the sum of demands of all the messages that was scheduled to time slice j. Let d' be the demand of the first message that was scheduled to time slice $j + 1$. Note that there is a message in time slice j whose demand is at least d', since the messages are sorted in non-increasing demand order. Therefore if $d' \geq C/2$, then $D_j \geq C/2$. If $d' < C/2$, then because this message was not assigned to time slice j we get that: $D_j > C - d' > C - C/2 = C/2$.

It follows that there is at most one time slot per color class that is not half full. Let ℓ denote the length of the schedule computed by NCMBL-APP. We obtain $\ell \leq \left\lceil \frac{n}{C/2} \right\rceil + \chi^* \leq 3 \cdot \ell^*$, and the claim follows.

4   Discussion and Open Problems

There are many open problems regarding smart antennas, and we outline here just a few. First, the assumption of unit arc-length may not hold for more complicated cases. Such cases may occur either due to the antenna's abilities or due to environmental conditions. Therefore, both capacitated models should be generalized to arbitrary circular-arc graphs.

We note that an approximation algorithm for capacitated coloring follows from an approximate coloring algorithm. Let ℓ^* denote the length of an optimal schedule and χ^* the chromatic number of the collision graph. By partitioning each color class into slots of size C and a possible leftover, it follows that $\ell^* \leq \lceil \frac{n}{C} \rceil + \chi^*$ (see the claim that bounds the approximation ratio). This implies also that a ρ-approximation for the chromatic number, leads to $(1 + \rho)$-approximation for capacitated coloring (with unit demands). In the case of non-unit demands, a ρ-approximation algorithm for coloring, leads to a $(2 + \rho)$-approximation algorithm.

The transmission model used in this paper is simple since we ignored physical effects such as reflections and side lobes. In order to take into account such effects, the model should be extended as follows. (i) Each message m_i defines a set of disjoint arcs. (ii) Two messages m_i, m_j can be transmitted concurrently only if every pair of arcs $a_i \in m_i$, $a_j \in m_j$ do not intersect (or if the amount of overlapping is limited). The collision graph that corresponds to this problem is more complicated, and finding an optimal schedule is an interesting open question even for the case where the number of arcs per message is constant.

Finally, improving the approximation factor for NU-C-MBL is interesting, since such an algorithm may be useful for more complicated models.

Acknowledgments

We thank Tony Weiss for interesting conversations about smart antennas. We thank the anonymous reviewers for the helpful remarks. Guy Even was partly funded by REMON, Israel 4G mobile consortium.

References

1. L. Epstein and A. Levin. On bin packing with conflicts. *Manuscript*, 2006.
2. M. R. Garey, D. S. Johnson, G. L. Miller, and C. H. Papadimitriou. The complexity of coloring circular arcs and chords. *SIAM Journal on Algebraic and Discrete Methods*, 1(2):216–227, June 1980.
3. F. Gavril. Intersection graphs of helly families of subtrees. *Discrete Appl. Math.*, 66(1):45–56, 1996.
4. K. Jansen. An approximation scheme for bin packing with conflicts. *J. Comb. Optim*, 3(4):363–377, 1999.
5. K. Jansen and S. Öhring. Approximation algorithms for time constrained scheduling. *Information and Computation*, 132(2):85–108, 1 Feb. 1997.
6. I. Karapetian. On coloring arc graphs. *Dokladi (Reports) of the Academy of Science of the Armenian Soviet Socialist Republic*, 70:306–311, 1980.
7. V. Kumar. An approximation algorithm for circular arc colouring. *ALGRTHMICA: Algorithmica*, 30, 2001.
8. M. Mouhamadou, P. Armand, P. Vaudon, and M. Rammal. Interference supression of the linear antenna arrays controlled by phase with use of sqp algorithm. *PIER: Progress In Electromagnetics Research*, 59:251–265, 2006.

9. J. B. Orlin, M. A. Bonuccelli, and D. P. Bovet. An $O(n^2)$ algorithm for coloring proper circular arc graphs. *SIAM Journal on Algebraic and Discrete Methods*, 2(2):88–93, June 1981.

10. W. K. Shih and W. L. Hsu. An approximation algorithm for coloring circular-arc graphs. In *SIAM Conference on Discrete Mathematics*, 1990.

11. A. Tucker. Coloring a family of circular arcs. *SIAM Journal on Applied Mathematics*, 29(3):493–502, Nov. 1975.

12. Valencia-Pabon. Revisiting tucker's algorithm to color circular arc graphs. *SICOMP: SIAM Journal on Computing*, 32, 2003.

13. P. K. Varlamos and C. N. Capsalis. Electronic beam steering using switched parasitic smart antenna arrays. *PIER: Progress In Electromagnetics Research*, 36:101–119, 2002.

14. D. B. West. *Introduction to Graph Theory (2nd Edition)*. (Prenctice Hall, Upper Saddle River), 2001.

15. Z. Zhang, M. Iskander, Z. Yun, and A. Host-Madsen. Hybrid smart antenna system using directional elements - performance analysis in flat rayleigh fading. *IEEE Transactions on Antennas and Propagation*, 51(10):2926–2935, 2003.

A   A Scheduling Algorithm for Big Color Classes

In this section we present an optimal scheduling algorithm for the case that $\left\lfloor \frac{n}{\chi^*} \right\rfloor > 2C$.

Definition 1. *Let C_i, C_j be two disjoint independent sets in a unit circular-arc graph. The sets $S_i \subseteq C_i$ and $S_j \subseteq C_j$ C-agree if: (i) $S_i \cup S_j$ is an independent set, and (ii) $|S_i \cup S_j| = C$.*

We now present an algorithm called BIG to handle the "big classes" case. We first compute an optimal balanced coloring by calling COL-PROPER(G). The algorithm BIG proceeds with $\chi^* - 1$ iterations. In the first iteration it does the following. (i) Partition the color class C_1 into C-groups, and a leftover group S_1. (ii) Schedule the C-groups. (iii) Find an independent set $S_2 \subseteq C_2$ that C-agrees with S_1. The existence of S_2 is proved in the next claim. Moreover, finding S_2 is easy: simply take a a subset of S_2 that are not neighbors of S_1. A listing of BIG appears as Algorithm 2. Since all slots, but perhaps the last slot, contain C arcs, Algorithm BIG computes a schedule whose length is $\left\lceil \frac{n}{C} \right\rceil$. Hence, we only need to prove the following claim.

Claim. There is always a set $S_{i+1} \subset C_{i+1}$ that C-agrees with the leftover subset of C_i.

Proof. Let S' and S'' denote two disjoint independent sets of arcs in a unit circular-arc graph G. We first show that if S'' is contained in the set $\Gamma(S')$ of neighbors of S', then $|S''| \leq 2 \cdot |S'|$. Assume otherwise, then there exists an arc $u \in S'$ that intersects at least 3 arcs $v_1, v_2, v_3 \in S''$. The arcs v_1, v_2, v_3 are independent. Therefore u must strictly contain one of these arcs (i.e., the middle arc). However, this contradicts the fact that all arcs are of unit length.

Algorithm 2. Algorithm BIG - handles the "big classes" case

1: Run COL-PROPER(G).
2: **for** $i = 1$ to $\chi^* - 1$ **do**
3:    Schedule the C-groups in what remains of C_i {*Scheduled messages are removed from C_i*}
4:    **if** $C_i \neq \emptyset$ **then**
5:       Find a set $S_{i+1} \subset C_{i+1}$ that C-agrees with C_i {*see claim below*}
6:       Schedule $S_{i+1} \cup C_i$, and remove S_{i+1} from C_{i+1}
7:    **end if**
8: **end for**
9: Schedule the C-groups and the leftover group in what remains of C_{χ^*}.

Consider the subset that remains of the color class C_i in Line 4. Let r denote the size of this leftover. By definition, $r < C$. By the previous paragraph, it follows that the number of arcs in C_{i+1} that intersect arcs in S_i is at most $2r$. Since $|C_{i+1}| - 2r > 2C - 2r > C - r$, we conclude that $|C_{i+1} \setminus \Gamma(S_i)| > C - r$. Hence, there are enough arcs in $C_{i+1} \setminus \Gamma(S_i)$ to be added to S_{i+1}.

We remark that the proof of the claim induces a linear time algorithm to find the required set S_{i+1}; simply add the first $C - t$ arcs that do not intersect S_i.

On Minimizing the Number of ADMs - Tight Bounds for an Algorithm Without Preprocessing

Michele Flammini[1], Mordechai Shalom[2], and Shmuel Zaks[2,*]

[1] Universita degli Studi dell'Aquila Dipartmento di Informatica, L'Aquila-Italy
flammini@di.univaq.it
[2] Department of Computer Science, Technion, Haifa-Israel
{cmshalom, zaks}@cs.technion.ac.il

Abstract. Minimizing the number of electronic switches in optical networks is a main research topic in recent studies. In such networks we assign colors to a given set of lightpaths. Thus the lightpaths are partitioned into cycles and paths. The switching cost is minimized when the number of paths is minimized. The problem of minimizing the switching cost is NP-hard, and approximation algorithms have been suggested for it. Many of these algoritms have a preprocessing stage, in which they first find cycles. The basic algorithm eliminates cycles of size at most l, and is known to have a performance guarantee of $OPT + \frac{1}{2}(1 + \epsilon)N$, where OPT is the cost of an optimal solution, N is the number of lightpaths, and $0 \leq \epsilon \leq \frac{1}{l+2}$, for any given odd l. Without preprocessing phase (i.e. $l = 1$), this reduces to $OPT + \frac{2}{3}N$. We develop a new technique for the analysis of the upper bound and prove a tight bound of $OPT + \frac{3}{5}N$ for the performance of this algorithm.

Keywords: Wavelength Assignment, Wavelength Division Multiplexing(WDM), Optical Networks, Add-Drop Multiplexer(ADM).

1 Introduction

1.1 Background

Given a WDM network $G = (V, E)$ comprising optical nodes and a set of full-duplex lightpaths $P = \{p_1, p_2, ..., p_N\}$ of G, the wavelength assignment (WLA) task is to assign a wavelength to each lightpath p_i.

In the following discussion we also assume that each lightpath $p \in P$ is contained in a cycle of G. Each lightpath p uses two ADM's, one at each endpoint. Although only the downstream ADM function is needed at one end and only the upstream ADM function is needed at the other end, full ADM's will be installed on both nodes in order to complete the protection path around some ring. The full configuration would result in a number of SONET rings. It follows that if

---

* This research was supported in part by the fund for the promotion of research at the Technion, by P. and E. Nathan research fund, and by the EU COST 293 (GRAAL) research fund.

T. Erlebach (Ed.): CAAN 2006, LNCS 4235, pp. 72–85, 2006.

two adjacent lightpaths are assigned the same wavelength, then they can be used by the same SONET ring and the ADM in the common node can be shared by them. This would save the cost of one ADM. An ADM may be shared by at most two lightpaths. A more detailed technical explanation can be found in [1].

Lightpaths sharing ADM's in a common endpoint can be thought as concatenated, so that they form longer paths or cycles. Each of these longer paths/cycles does not use any edge $e \in E$ twice, for, otherwise they cannot use the same wavelength and this is a necessary condition to share ADM's.

1.2 Previous Work

Minimizing the number of electronic switches in optical networks is a main research topic in recent studies. The problem was introduced in [1] for ring topology. An approximation algorithm for ring topology with approximation ratio of 3/2 was presented in [2], and was improved in [3,4] to $10/7 + \epsilon$ and $10/7$, respectively.

For general topology [5] describes an algorithm with approximation ratio of 1.6. The same problem was studied in [6] and two algorithms are proposed for the solution: The $MCC - WS$ algorithm is proven to have a performance of $OPT + \frac{3}{5}N$ where OPT is the cost of an optimal solution, and N is the number of lightpaths. In the same work another algorithm (which we name $PIM(l)$) with a preprocessing phase where cycles of length at most l are included in the solution is shown to have performance guarantee of

$$OPT + \frac{1}{2}(1 + \epsilon)N, \quad 0 \le \epsilon \le \frac{1}{l+2} \tag{1}$$

for any given odd l. This result was improved to $\frac{1}{2l+3} \le \epsilon \le \frac{1}{\frac{3}{2}(l+2)}$ in [7]. Substituting $l = 1$, gives raise to an algorithm without preprocessing, namely $IM = PIM(1)$, with performance of $OPT + \frac{2}{3}N$ according to [6] and $OPT + \frac{11}{18}N$ according to [7].

1.3 Our Contribution

In this work, using a novel technique, we improve the analysis of algorithm IM. We prove a tight bound of $OPT + \frac{3}{5}N$ for the performance of IM, namely

$$OPT + \frac{1}{2}(1 + \epsilon)N, \quad \epsilon = \frac{1}{5}. \tag{2}$$

An interesting fact is that this bound matches exactly the performances proven for the EMZ algorithm in [5] and the performance of algorithm $MCC - WS$ in [6].

In Section 2 we describe the problem, in Section 3 we introduce some preliminary results. The algorithm is presented in Section 4. In Sections 5 and 6 we present our upper and lower bounds, respectively.

2   Problem Definition

An instance α of the problem is a pair $\alpha = (G, P)$ where $G = (V, E)$ is an undirected graph and P is a set of simple paths in G. Given such an instance we define the following:

Definition 2.1. *The paths $p, p' \in P$ are conflicting or overlapping if they have an edge in common. This is denoted as $p \asymp p'$. The graph of the relation $\asymp$ is called the conflict graph of (G, P).*

Definition 2.2. *A proper coloring (or wavelength assignment) of P is a function $w : P \mapsto \mathbb{N}$, such that $w(p) \neq w(p')$ whenever $p \asymp p'$.*

Note that w is a proper coloring if and only if for any color $c \in \mathbb{N}$, $w^{-1}(c)$ is an independent set in the conflict graph.

Definition 2.3. *A valid chain (resp. cycle) is a path (resp.cycle) formed by the concatenation of distinct paths $p_0, p_1, ..., p_{k-1} \in P$ that do not go over the same edge twice. Note that the paths of a valid chain (resp. cycle) constitute an independent set of the conflict graph.*

Definition 2.4. *A solution S of an instance $\alpha = (G, P)$ is a set of chains and cycles of P such that each $p \in P$ appears in exactly one of these sets.*

In the sequel we introduce the shareability graph, which together with the conflict graph constitutes another (dual) representation of the instance α. Except for one particular case, we will use the dual representation of the problem.

Definition 2.5. *The shareability graph of an instance $\alpha = (G, P)$, is the edge-labelled multi-graph $\mathcal{G}_\alpha = (P, E_\alpha)$ such that there is an edge $e = (p, q)$ labelled u in E_α if and only if $p \not\asymp q$, and u is a common endpoint of p and q in G.*

Example: Let $\alpha = (G, P)$ be the instance in Figure 1. Its shareability graph $\mathcal{G}_\alpha$ is the graph at the left side of Figure 2.

In this instance $P = \{a, b, c, d, e\}$, and it constitutes the set of nodes of $\mathcal{G}_\alpha$. The edge set is $E_\alpha = \{(b, c, u), (d, e, v), (a, c, w), (a, b, x), (a, d, x), (b, e, x), (d, e, x)\}$, because b and c can be joined in their common endpoint u, etc.. Note that, for instance $(b, d, x) \notin E_\alpha$, because although they share a common endpoint x, b and d can not be concatenated, as they have the edge (x, u) in common. The corresponding conflict graph is the graph at the right side of Figure 2. It has the same node set and the edge set is $\{(c, d), (b, d), (c, e), (a, e)\}$. The paths $c, d \in P$ are conflicting because they have a common edge, i.e. (u, v), etc.

Note that the edges of the conflict graph are not in E_α. This immediately follows from the definitions.

Note also that, for any node v of $\mathcal{G}_\alpha$, the set of labels of the edges adjacent to v is of size at most two.

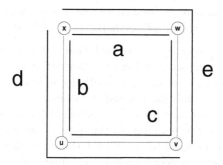

Fig. 1. A sample input

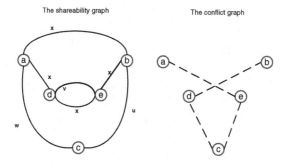

Fig. 2. The shareability and conflict graphs

Definition 2.6. *A valid chain (resp. cycle) of $\mathcal{G}_\alpha$ is a simple path $p_0, p_1, ..., p_{k-1}$ of $\mathcal{G}_\alpha$, such that any two consecutive edges in the path (resp. cycle) have distinct labels and its node set is properly colorable with one color (in G), or in other words constitutes an independent set of the conflict graph.*

Note that the valid chains and cycles of $\mathcal{G}_\alpha$ correspond to valid chains and cycles of the instance. In the above example the chain a, d which is the concatenation of the paths a and d in the graph G, corresponds to the simple path a, d in $\mathcal{G}_\alpha$ and the cycle a, b, c which is a cycle formed by the concatenation of three paths in G corresponds to the cycle a, b, c in $\mathcal{G}_\alpha$. Note that no two consecutive labels are equal in this cycle. On the other hand the paths b, a, d can not be concatenated to form a chain, because this would require the connection of a to both b and d at node x. The corresponding path b, a, d in $\mathcal{G}_\alpha$ is not a chain because the edges (b, a) and (a, d) have the same label, namely x.

Definition 2.7. *The sharing graph of a solution S of an instance $\alpha = (G, P)$, is the following subgraph $\mathcal{G}_{\alpha,S} = (P, E_S)$ of $\mathcal{G}_\alpha$. Two lightpaths $p, q \in P$ are connected with an edge labelled u in E_S if and only if they are consecutive in a* *chain or cycle in the solution S, and their common endpoint is $u \in V$. We will usually omit the index α and simply write $\mathcal{G}_S$. $d(p)$ is the degree of node p in $\mathcal{G}_S$.*

In our example, $S = \{(a,d)(b,e),(c)\}$ is a solution with three chains. The sharing graph of this solution is depicted in Figure 3. Note that for a solution consisting of chains of size at most two, the distinct labelling condition is satisfied vacuously, and the independent set condition is satisfied because no edge of $\mathcal{G}_\alpha$ can be edge of the conflict graph. Another possible solution is S' consisting of the cycles (a,b,c) and (d,e), the corresponding sharing graph contains two cycles (a,b,c) and (d,e). In this case we need to check that the remaining two conditions are satisfied. Indeed, no two consecutive labels are equal and each cycle constitutes an independent set of the conflict graph.

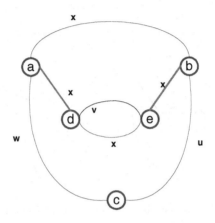

Fig. 3. A possible solution

We define:
$$\forall i \in \{0,1,2\}, D_i(S) \overset{def}{=} \{p \in P | d(p) = i\}$$

and

$$d_i(S) \overset{def}{=} |D_i(S)|.$$

Note that $d_0(S) + d_1(S) + d_2(S) = |P| = N$.

An edge $(p,q) \in E_S$ with label u corresponds to a concatenation of two paths with the same color at their common endpoint u. Therefore these two endpoints can share an ADM operating at node u, thus saving one ADM. We conclude that every edge of E_S corresponds to a saving of one ADM. When no ADMs are shared, each path needs two ADM's, for a total of $2N$ ADMs. Therefore the cost of a solution S is

$$cost(S) = 2|P| - |E_S| = 2N - |E_S|.$$

The objective is to find a solution S such that $cost(S)$ is minimum, in other words $|E_S|$ is maximum.

3 Preliminary Results

In this section we observe some basic properties of solutions which are independent of the algorithms used.

Given a solution S, $d(p) \leq 2$ for every node $p \in P$. Therefore, the connected components of $\mathcal{G}_S$ are either paths or cycles. Note that an isolated vertex is a special case of a path. Let $\mathcal{P}_S$ be the set of the connected components of $\mathcal{G}_S$ that are paths. Clearly, $|E_S| = N - |\mathcal{P}_S|$. Therefore

$$cost(S) = 2N - |E_S| = N + |\mathcal{P}_S|$$

Let S^* be a solution with minimum cost. For any solution S we define

$$\epsilon(S) \stackrel{def}{=} \frac{d_0(S) - d_2(S) - 2\,|\mathcal{P}_{S^*}|}{N}.$$

Lemma 1. *For any solution* S

$$cost(S) = cost(S^*) + \frac{1}{2}N(1 + \epsilon(S)).$$

Proof. Clearly $|E_{S^*}| = N - |\mathcal{P}_{S^*}|$. On the other hand $2\,|E_S|$ is the sum of the degrees of the nodes in $\mathcal{G}_S$, namely

$$2\,|E_S| = d_1(S) + 2d_2(S) = N - d_0(S) + d_2(S)$$

We conclude:

$$
\begin{aligned}
cost(S) - cost(S^*) = |E_{S^*}| - |E_S| &= N - |\mathcal{P}_{S^*}| - \frac{N - d_0(S) + d_2(S)}{2} \\
&= \frac{N}{2} + \frac{d_0(S) - d_2(S) - 2\,|\mathcal{P}_{S^*}|}{2} \\
&= \frac{1}{2}N\left(1 + \frac{d_0(S) - d_2(S) - 2\,|\mathcal{P}_{S^*}|}{N}\right)
\end{aligned}
$$

$\square$

The following definition extends the concept of a chord from cycles to paths.

Definition 3.1. *Given an instance* $\alpha = (G, P)$ *and a solution* S *of* α, *an edge* (p, q) *of* $\mathcal{G}_\alpha$ *is a chord of* S *if both* p *and* q *are in the same connected component of* $\mathcal{G}_S$ *and* $(p, q) \notin E_S$.

Lemma 2. *For every instance* $\alpha = (G, P)$ *there is an optimal solution* S^* *without chords.*

Proof. Note that any two solutions S_1, S_2 of α such that $cost(S_1) = cost(S_2)$, have the same number of chains, whereas the number of cycles may differ. Let S^* be a solution with maximum number of cycles among the solutions with minimum cost, i.e. optimal. We will prove that S^* satisfies the claim.

We claim that there is no node $v \in V$ and no chain (resp. cycle) C of S^*, such that v is used more than once as an endpoint of a paths in C. Assume the contrary. Consider two occurrences of v in C. It is impossible that C is a path and v terminates both ends C. In this case C can be closed to a cycle, and get a solution with one path less, contradicting the optimality of S^*. Consider the sequence of paths between these two occurrences of v. This is a valid cycle, say C'. Consider the solution S' obtained by taking S^* and separating C into two parts. The first part is C' and the second part is the sequence obtained by the concatenation of the paths before the first occurrence of v with the paths after the second occurrence of v, where one of these but not both may be empty. S' has the same number of paths as S^*, therefore $cost(S') = cost(S^*)$, therefore optimal. Moreover, S' has one more cycle than S^*, contradictory to the way S^* was chosen.

Assume that (p, q) is a chord of S^*. Let x be its label. Then x is an endpoint of both p and q. Because (p, q) is a chord, $(p, q) \notin E_{S^*}$, in other words p and q do not have the node x as common endpoint in this connected component. Then x appears at least twice in the connected component, a contradiction. Therefore there are no chords of S^*. $\square$

4   Algorithm IM

In this section we give a short description of the algorithm in [2], without preprocessing. The algorithm begins with chains consisting of single nodes [1] (which are always valid). At each iteration, we try to combine a maximum number of pairs of chains to obtain longer chains (in fact, less chains). This is done by constructing an appropriate graph and computing a maximum matching on it. The algorithm ends when the maximum matching is empty, namely no two chains can be combined to a longer chain.

```
Phase 0) E_S = ∅
         // the chains of G_S are isolated nodes.
Phase 1)      Do {
         Build the graph G'_α in which each node is a chain of G_S
         and there is an edge labelled u between two chains if and
         only if the chains can be merged into one bigger chain
         by joining them at a common endpoint u.
         //In the first iteration G'_α = G_α
         Find a maximum matching MM of G'_α.
         For each edge e = (c, c') of MM labelled u do {
             Merge the corresponding chains into one chain
             by joining them in the common endpoint u
             // Note that the chains may have an additional
             // endpoint, say v, which is not affected
         }
       } Until MM = ∅.
```

---

[1] We use the dual representation, in which an element p of P is referred as a node (of G_α), and a path refers to a path of G_α.

For completeness, we briefly argue about the correctness of the algorithm: After Phase 0, the chains of S consist of single nodes. Trivially, these are valid chains. At each iteration of Phase 1, a new chain is constructed only if it is valid, because edges are added to $\mathcal{G}'_\alpha$ only if the corresponding chains can be merged into one chain. Each edge of a matching represents a valid merging operation. Moreover two such valid operations do not affect each other, because each such operation is performed on two chains matched by an edge of some matching. Therefore after each iteration the solution consists of valid chains.

5   Upper Bound

We will modify the algorithm IM so that its performance can be only worse and then analyze a solution returned by the modified algorithm. We make two modifications:

- The algorithm performs only two iterations.
- In the second iteration instead of a maximum matching the algorithm finds a maximal bipartite matching where one set are isolated nodes of $\mathcal{G}_S$ and and the second set of nodes are paths of length one in $\mathcal{G}_S$.

After the first iteration $\mathcal{G}_S$ contains isolated nodes and paths of length one. After the second iteration $\mathcal{G}_S$ contains paths of length at most two.

In the sequel S is a solution returned by the modified algorithm and S^* is an optimal solution without chords, whose existence is guaranteed by Lemma 2.

We direct each edge of $\mathcal{G}_{S^*}$, such that each path becomes a directed path and each cycle becomes a directed cycle. The direction chosen for every path (resp. cycle) is arbitrary. Let $\overrightarrow{\mathcal{G}}_{S^*}$ be the digraph obtained by this process. Unless otherwise stated, $d_{in}(p)$ and $d_{out}(p)$, denote the in and out degrees of p in $\overrightarrow{\mathcal{G}}_{S^*}$, respectively. Clearly, $\forall p \in P$, $d_{in}(p) \leq 1$ and $d_{out}(p) \leq 1$. The following definitions refer to $\overrightarrow{\mathcal{G}}_{S^*}$:

$LAST^*$ is the set of nodes that do not have successors in $\overrightarrow{\mathcal{G}}_{S^*}$, namely

$$LAST^* \stackrel{def}{=} \{p \in P | d_{out}(p) = 0\}.$$

Note that $|LAST^*| = |P_{S^*}|$.

The functions $Next^*$ and $Prev^*$ are defined as expected: $Next^*$ (resp. $Prev^*$) maps a node p to the next (resp. previous) node in $\overrightarrow{\mathcal{G}}_{S^*}$ whenever such a node exists, namely:

$$Next^* : P \setminus LAST^* \mapsto P$$

and $Next^*(p)$ is the unique node u such that there is an edge from p to u in $\overrightarrow{\mathcal{G}}_{S^*}$. $Prev^* = Next^{*-1}$.

Let MM be the maximum matching found by the algorithm in the first iteration of Phase 1. We make the following observation:

Observation 5.1

- **a)** *An edge $e = (p, q) \in E_S$ such that $d(p) = d(q) = 1$ is in MM.*
- **b)** *Let p, q, r be a maximal path of $\mathcal{G}_S$. We can assume that either $e = (p, q) \in MM$ and $e' = (q, r) \notin MM$ or vice versa.*

Proof. — **a)** Assume $e \notin MM$, then at the end of the first iteration $d(p) = d(q) = 0$. This implies that MM is not a maximum matching, a contradiction.
- **b)** Obviously, either e or e' is in MM. Otherwise one of them can be added to MM and augment it. Assume $e \in MM$, then $e' \notin MM$. $MM' = MM - \{e\} \cup \{e'\}$ is a maximum matching too. As the algorithm may return any maximum matching in its first phase we may equally assume that MM' is the matching returned in the first phase and e is added to the solution in a subsequent phase. □

Using the notation in Lemma 1, we will prove

Theorem 1. *For any solution S returned by algorithm IM*

$$cost(S) \leq cost(S^*) + \frac{3}{5}N.$$

Proof. We partition $D_0(S)$ into the sets A, B, C and D using the following classification procedure **CLASSIFY**:

Given $p \in D_0(S)$, **CLASSIFY** finds a sequence $f(p) = (p_0, p_1, ...)$ of elements of P.

CLASSIFY $(p \in P)$ {

- $p_0 = p$
- For $i \geq 1$ do:
    - a)If $p_{i-1} \in D_2(S)$ then $p \in A$, $f(p) = (p_0, ..., p_{i-1})$, **return**.
    - b)If $p_{i-1} \in LAST^*$ then $p \in B$, $f(p) = (p_0, ..., p_{i-1})$, **return**.
    - c)If there is a node repeated at least twice in the sequence $p_0, ..., p_{i-1}$ then $p \in C$, $f(p) = (p_0, ..., p_{i-1})$, **return**.
    - d)If i is even:

        If $p_{i-1} \in D_1(S)$ then
            p_i is the (unique) neighbor of p_{i-1} in $\mathcal{G}_S$,
        else /* $p_{i-1} \in D_0(S)$ */
            $p \in D$, $f(p) = (p_0, ..., p_{i-1})$, **return**.
    - e-If i is odd:
        $p_i = Next^*(p_{i-1})$. // note that this is always possible because $p_{i-1} \notin LAST^*$.

}

Clearly, the above procedure terminates, with a finite sequence $f(p) = (p_0, ..., p_{i-1})$. This is because in each iteration of the loop, either the procedure ends, or a node is added to the sequence. P is finite, therefore, eventually a node will be added twice to the sequence, unless we terminate earlier. Whenever this happens the procedure terminates in the next iteration.

We define the following sets which will be useful in the sequel.

$$E_{odd} = \{(p_j, p_j + 1) \in E_\alpha | j \text{ is odd} \}$$

$$E_{even} = \{(p_j, p_j + 1) \in E_\alpha | j \text{ is even} \}.$$

Note that nodes are added to $f(p)$ only in steps **d)** and **e)**. By inspection of the code of these steps we conclude $E_{odd} \subseteq E_S$ and $E_{even} \subseteq E_{S^*}$.

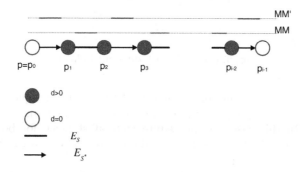

Fig. 4. Case D is impossible

Claim. $D = \emptyset$.

Proof. Assume $p \in D$. Then the classification procedure $CLASSIFY(p)$ ends with a sequence $p = p_0, ..., p_{i-1}$ such that $p_{i-1} \in D_0(S)$ (see Figure 4). p_0 and p_{i-1} are isolated vertices of $\mathcal{G}_S$ and the other nodes are of degree one. Then $E_{odd} \subseteq MM \subseteq E_S$. Note also that i is even. Then $MM' = MM \setminus E_{odd} \cup E_{even}$ is a matching such that $|MM'| = |MM| + 1$; in other words $p_0, ..., p_{i-1}$ is an augmenting path for the maximum matching MM, a contradiction.

Claim. The sets $f(p)$ are pairwise disjoint.

Proof. Assume by contradiction that $p \neq q$ and $f(p) \cap f(q) \neq \emptyset$. Let r be the first element of $f(p)$ in the intersection. Recall that $d(r) \in \{0, 1, 2\}$. We consider the three cases separately:

- $d(r) = 0$: $p, q \in D$ is impossible by Claim 5. Therefore a node with degree 0 may appear only as the first node. Then, the only degree 0 node in $f(p)$ is p and the only degree 0 node in $f(q)$ is q. Then $p = r = q$, a contradiction.
- $d(r) = 1$: We divide this case into subcases:

- r has odd index in one of the sequences (say $f(p)$) and even index in the other(say $f(q)$). In this case the path $(p = p_0, p_1, ..., r, ..., q_1, q_0 = q)$ is an augmenting path for MM, a contradiction. (See figure 5).
- r has odd indices in both sequences. In this case $prev(r) \in f(p) \cap f(q)$, and has an index lower than r in $f(p)$, a contradiction.

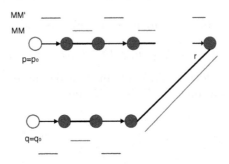

Fig. 5. $d(r) = 1$, with different indices

- r has even indices in both sequences. Let r' be the unique neighbor of r in $\mathcal{G}_S$. Then $r' \in f(p) \cap f(q)$ and occurs before r in $f(p)$, a contradiction.
- $d(r) = 2$: In this case, the procedure ends at step **a)**, for both p and q. Therefore $p, q \in A$. Let r' and r'' be the neighbors of r in $\mathcal{G}_S$. We consider three subcases as before:
  - r has odd index in one of the sequences (say $f(p)$) and even index in the other(say $f(q)$). One of r', r'' is in $f(q)$. Without loss of generality assume $r' \in f(q)$. Then $d(r') = 1$. Therefore (r', r, r'') is a maximal path in $\mathcal{G}_S$ (see Figure 6). By Observation 5.1 we may assume $(r, r') \in MM$ and $(r, r'') \notin MM$. Then $q = q_0, q_1, ..., r', r, ..., p_1, p_0 = p$ is an augmenting path for MM, a contradiction.
  - r has odd indices in both sequences. In this case $prev(r) \in f(p) \cap f(q)$, and has an index lower than r in $f(p)$, a contradiction.
  - r has even indices in both sequences (see Figure 7). Then without loss of generality $r' \in f(p)$ and $r'' \in f(q)$. Therefore $d(r') = d(r'') = 1$. Therefore (r', r, r'') is a maximal path in $\mathcal{G}_S$. We may assume $(r, r') \in MM$ and $(r, r'') \notin MM$. Then the path $q = q_0 q_1 ... r''$, is an augmenting path, a contradiction.

Claim. If $p \in C$ then $|f(p)| \geq 5$.

Proof. Let $p \in C$. Then, the node p_{i-1} is the first node repeated twice in the sequence, namely $\exists j \leq i - 2$ such that $p_j = p_{i-1}$. First, we will prove that i is even: Assume i is odd, then $(p_{i-2}, p_{i-1}) = (p_{i-2}, p_j) \in E_{odd}$ (see Figure 8). If $j = 0$ then $d(p) = d(p_0) \geq 1$, a contradiction. If $j > 0$ then $d(p_j) \geq 2$, in this case the procedure would have stopped earlier in step **a)**, and then $p \in A$, a contradiction.

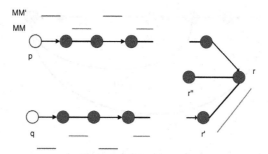

Fig. 6. $d(r) = 2$ with different indices

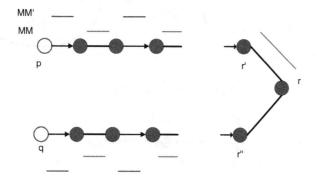

Fig. 7. $d(r) = 2$ with even indices

Therefore $|f(p)| = |\{p_0, ..., p_{i-2}\}| = i - 1$ is odd. If $j = 0$ then $f(p)$ is a cycle of $\mathcal{G}_\alpha$, otherwise $f(p)$ contains a cycle of $\mathcal{G}_\alpha$, denoted by $c(p)$. In the latter case $f(p)$ is called a spoon and $h(p) \overset{def}{=} f(p) \setminus c(p)$ is the handle of the spoon (see Figure 9). $|f(p)| > 1$, because a self loop in $\mathcal{G}_\alpha$ can not be a simple path of G. Assume by contradiction that $|f(p)| = 3$. Then $f(p)$ is a cycle $\{p, q, r\}$ (see figure 10). In this case $u \neq v$, for rpq is in a connected component of $\mathcal{G}_{S^*}$. Then x is different from at least one of u, v. Assume w.l.o.g $x \neq u$. Then (p, q) could be added to the E_S in the second iteration. Therefore $d(p) \geq 1$, a contradiction.

We claim that $|c(p)| \geq 5$. Assume, by contradiction that $c(p)$ is the cycle $\{p, q, r\}$ as in Figure 10. Then we consider two cases. Either the edge (q, r) labelled x is a chord, a contradiction, or it is in E_{S^*}, then $c(p)$ is a cycle of the optimal solution. In this case one of the edges $(p, q), (p, r)$ would be added to E_S in the next iteration of the algorithm. We conclude $|f(p)| \geq |c(p)| \geq 5$.

Now we complete the proof of the lemma:

- For $p \in A$, $f(p)$ contains exactly one node $p' \neq p$ from $D_2(S)$. Therefore $|A| \leq |D_2(S)| = d_2(S)$.

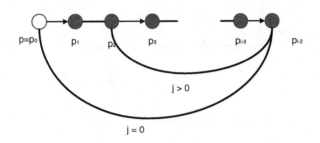

Fig. 8. i cannot be even

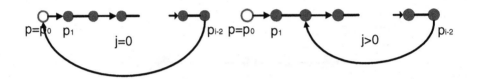

Fig. 9. A Cycle and a Spoon

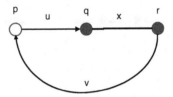

Fig. 10. A cycle of length 3

- For $p \in B$, $f(p)$ contains exactly one node from $LAST^*$. Therefore $|B| \leq |LAST^*|$.
- For $p \in C$, $f(p)$ contains at least 5 nodes, therefore $|C| \leq N/5$.

$D_0(S) = A \uplus B \uplus C \uplus D$, then:

$$d_0(S) = |A| + |B| + |C| + |D| \leq d_2(S) + |LAST^*| + N/5$$

$$\epsilon(S) = \frac{d_0(S) - d_2(S) - 2|P_{S^*}|}{N} \leq \frac{d_0(S) - d_2(S) - |LAST^*|}{N} \leq \frac{1}{5}.$$

This combined with Lemma 1, completes the proof. ☐

6   Lower Bound

Theorem 2. *There are infinitely many instances (G, P) and solutions S returned by $PIM(1)$, such that*

$$cost(S) = cost(S^*) + (3/5)N.$$

Proof. For each natural number k, consider the instance which is obtained by duplicating k times the instance depicted in Figure 1. The algorithm may return k copies of the solution depicted in Figure 3, because the matching consisting of the edges $\{(a, d), (b, e)\}$ is a maximum matching. If the algorithm finds this maximum matching in the first iteration, it will not be able to extend it in any manner in the next phase and the algorithm will terminate $\mathcal{G}_S$ being this maximum matching. We therefore have

$$d_0(S) = k, d_2(S) = 0, |\mathcal{P}_{S^*}| = 0, N = 5k$$

and

$$\epsilon(S) = \frac{d_0(S) - d_2(S) - 2\,|\mathcal{P}_{S^*}|}{N} = \frac{1}{5}.$$

This, together with Lemma 1 implies the claim.   □

References

1. O. Gerstel, P. Lin, and G. Sasaki. Wavelength assignment in a wdm ring to minimize cost of embedded sonet rings. In *INFOCOM'98, Seventeenth Annual Joint Conference of the IEEE Computer and Communications Societies*, pages 69–77, 1998.
2. G. Călinescu and P-J. Wan. Traffic partition in wdm/sonet rings to minimize sonet adms. *Journal of Combinatorial Optimization*, 6(4):425–453, 2002.
3. M. Shalom and S. Zaks. A $10/7 + \epsilon$ approximation scheme for minimizing the number of adms in sonet rings. In *First Annual International Conference on Broadband Networks, San-José, California, USA*, October 2004.
4. L. Epstein and A. Levin. Better bounds for minimizing sonet adms. In *2nd Workshop on Approximation and Online Algorithms, Bergen, Norway*, September 2004.
5. T. Eilam, S. Moran, and S. Zaks. Lightpath arrangement in survivable rings to minimize the switching cost. *IEEE Journal of Selected Area on Communications*, 20(1):172–182, Jan 2002.
6. G. Călinescu, O. Frieder, and P.-J. Wan. Minimizing electronic line terminals for automatic ring protection in general wdm optical networks. *IEEE Journal of Selected Area on Communications*, 20(1):183–189, Jan 2002.
7. M. Flammini, M. Shalom, and S. Zaks. On minimizing the number of adms in a general topology optical network. *Accepted to the 20th International Symposium on Distributed Computing (DISC), Stockholm, Sweden*, September 2006.

Tolerance Based Contract-or-Patch Heuristic for the Asymmetric TSP

Boris Goldengorin[1,2], Gerold Jäger[3], and Paul Molitor[3]

[1] Faculty of Economic Sciences, University of Groningen,
9700 AV Groningen, The Netherlands
B.Goldengorin@rug.nl
[2] Department of Applied Mathematics,
Khmelnitsky National University, Ukraine
[3] Computer Science Institute, University of Halle-Wittenberg,
D-06099 Halle (Saale), Germany
jaegerg@informatik.uni-halle.de,
paul.molitor@informatik.uni-halle.de

Abstract. In this paper we improve the quality of a recently suggested class of construction heuristics for the Asymmetric Traveling Salesman Problem (ATSP), namely the Contract-or-Patch heuristic. Our improvement is based on replacing the selection of each path to be contracted after deleting a heaviest arc from each short cycle in an Optimal Assignment Problem Solution (OAPS) by contracting a single arc from a short cycle in an OAPS with the largest upper tolerance with respect to one of the relaxed ATSP. The improved algorithm produces higher-quality tours than all previous COP versions and is clearly outperforming all other construction heuristics on robustness.

Keywords: Traveling Salesman Problem, Tolerances, Construction Heuristics.

1 Introduction

The Traveling Salesman Problem (TSP) is the problem of finding a shortest tour through a given number n of locations such that every location is visited exactly once. The cost of traveling from location i to location j is denoted by $c(i,j)$. These costs are called *symmetric* if $c(i,j) = c(j,i)$ holds for each pair of cities i and j, and *asymmetric* otherwise. If the costs are symmetric, we refer to the TSP as *symmetric TSP* (STSP) and if the costs are asymmetric, as *asymmetric TSP* (ATSP). A TSP instance is defined by the entries of an $n \times n$ matrix $C = ||c(i,j)||$.

Tour construction heuristics build a tour without attempting to improve it once it is constructed. They can be used to provide starting tours for local search and branch and bound algorithms or to find approximate solutions of the ATSP when little time is available. It is important to note that heuristics of this type are usually very fast and therefore are positioned differently from

T. Erlebach (Ed.): CAAN 2006, LNCS 4235, pp. 86–97, 2006.

more expensive ATSP approximation techniques such as truncated branch-and-bound [18] and improvement heuristics [12]. Glover et al. [2] have proposed three new tour construction heuristics, one of which, Contract-or-Patch (denoted by COP), has been shown to be a robust heuristic for a variety of classes of ATSP instances. COP combines modifications of two other heuristics, Recursive Path Contraction (RPC) [2], and Greedy Karp-Steele Patching (GKS), denoted by COP/GKS. A poor feature of COP/GKS is its considerable programming complexity which implies comparatively high computation complexity. The request to reduce this complexity has motivated Gutin and Zverovich [9] to analyze comparative performance of COP/GKS and a combination of COP with the simpler Karp-Steele Patching (KSP) algorithm, denoted by COP/KSP. Note that both heuristics, namely COP/GKS and COP/KSP perform very similarly in terms of the quality of tours they produce and their running times. The simplicity of COP/KSP makes it a good candidate for replacing COP/GKS in most, if not all, potential applications. Based on a computational study of all possible COP versions, Gutin and Zverovich [9] have introduced an expansion of contracted arcs and have shown that two updated heuristics KSP/COP and GKS/COP produce higher quality tours compared to the original COP/KSP and COP/GKS procedures. All of the above mentioned heuristics are contracting all remaining paths after deleting a heaviest arc from each short cycle in an optimal assignment problem solution and include all paths into the unknown heuristic solution. A drawback of this strategy is that costs of deleted arcs are no accurate indicators whether those arcs are excluded from a "good" TSP solution. In [6] and [18], it is shown that *tolerances* are better indicators. An upper (lower) tolerance of an arc is the cost of excluding (including) that arc from (in) the optimal solution at hand. Although the concept of tolerances has been applied for decades (in sensitivity analysis; see e.g. [7] and [16]), only Helsgaun's version of the Lin-Kernighan heuristic for the STSP applies tolerances (see [11]).

In this paper we introduce an improved version of the GKS/COP algorithm, which applies a criterion based on approximated upper tolerances for selecting the arcs to be contracted instead of the costs based criterion and keep the GKS unchanged. We perform a computational evaluation of this new heuristic denoted by GKS/TBCOP, and show that it produces higher quality tours compared to the original KSP/COP procedure. We also perform an extensive computational study of the impact of a COP parameter, the threshold, on tour quality and running time of both versions of COP and TBCOP, and recommend a robust choice for the value of the threshold. The evaluation is performed on a diverse set of ATSP instances, comprising nine different families of instances of which seven are exactly the same as in Glover et al. [2] and Gutin and Zverovich [9].

Our paper is organized as follows. In the next section we briefly describe the Contract-or-Patch algorithms COP/KSP, COP/GKS, KSP/COP, and GKS/COP introduced in [2] and [9]. In Section 3 we present our tolerance based version of the Contract-or-Patch heuristic. The results of computational experiments are reported in Section 4. The conclusions and future research directions are discussed in Section 5.

2 Contract-or-Patch Heuristics

The Contract-or-Patch class of heuristics suggested in Glover et al. [2] and recently improved in Gutin and Zverovich [9] is based on two operations: contraction and patching after solving the Assignment Problem (AP) as a relaxation of the ATSP. (For undefined graph theory terminology see e.g., Bang-Jensen and Gutin [1].)

A feasible solution of the AP is called a *cycle cover*, an optimal solution with respect to cost function c *minimum cycle cover*. A minimum cycle cover consists of a number of cycles $F = \{C_1, \ldots, C_k\}$ and can be computed in $O(n^3)$ with the Hungarian method (see e.g. [13]). The length of a cycle $C_p = \{a_1, \ldots, a_q\}$ for each $p = 1, \ldots, k$ is the number q of arcs a_l with $l = 1, \ldots, q$ included in the cycle C_p. In terms of the ATSP a feasible AP solution requires that each city will be visited exactly once without necessarily creating a single (*Hamiltonian*) cycle. The purpose of COP heuristics is to transform a minimum cycle cover into a Hamiltonian cycle (tour) with its total costs as close as possible to the shortest Hamiltonian cycle by recursive application of three operations, namely the contraction, recontraction (expansion of all contracted arcs), and patching.

Let us define all of these three operations in terms of the weights (costs) of the given complete weighted directed graph $G = (V(G), A(G), c)$ with the vertex set $V(G) = \{1, \ldots, n\}$, arc set $A(G) = \{(i, j) : i, j \in V(G)\}$, weight (cost) function $c = c(i, j)$ for all $(i, j) \in A(G)$, and minimum cycle cover $F = \{C_1, \ldots, C_k\}$ with $k > 1$ and $2 \le |C_p| \le n - 2$ for each $p = 1, \ldots, k$ and $\sum_{p=1}^{k} |C_p| = n$.

Contraction: Given the directed path $P = v_1 v_2 \ldots v_s$ with $2 \le s \le n - 3$ obtained after the deletion of a single arc from the cycle C_p. The operation of contraction of P leads to the new graph G_P with $V(G_P) = (V(G) \cup \{p\}) \setminus \{v_1, v_2, \ldots, v_s\}$, where p is a new vertex. The weight $c_P(x, y)$ of an arc (x, y) of the graph G_P is defined as

$$c_P(x, y) = \begin{cases} c(x, y) & \text{if } x \ne p \text{ and } y \ne p \\ c(v_s, y) & \text{if } x = p \text{ and } y \ne p \\ c(x, v_1) & \text{if } x \ne p \text{ and } y = p \end{cases}$$

The special case of contracting an arc (x, y) means contracting the path xy of length one.

Recontraction (expansion): Given a contracted graph $G_C = (V(G_C), A(G_C))$ with $F = \{C_1, \ldots, C_k\}$ and the set of new (recursively contracted) vertices $p_1, \ldots, p_t$ included in F. The operation of recontraction (expansion) of F recursively replaces each contracted vertex p_l for all $l = 1, \ldots, t$ by its path from the original graph G leading to a new cycle cover with the same number k of cycles.

Patching: The operation of patching two cycles C_i and C_j from F with $i \ne j$ is defined as follows: two fixed arcs $(x_i, y_i) \in C_i$ and $(x_j, y_j) \in C_j$ are deleted and two arcs (x_i, y_j) and (x_j, y_i) joining the cycles together are added. The cost of patching C_i and C_j using (x_i, y_j) and (x_j, y_i) is defined as $\delta(C_i, C_j) = c(x_i, y_j) + c(x_j, y_i) - (c(x_i, y_i) + c(x_j, y_j))$, i.e., $\delta(C_i, C_j)$ is the difference between the sum of the costs of the inserted arcs and the sum of the costs of the deleted arcs. The following two versions of patching algorithms are introduced in [2], the KSP (Karp-Steele-Patching) (see also [14], [15]) and the GKS (Greedy Karp-Steele-Patching).

Algorithm 1 (KSP)

1 *Solve the AP with a minimum cycle cover F.*
2 *Patch two longest cycles (i.e. cycles with most vertices) of F (the two arcs are chosen in such a way that the patching costs become minimum).*
3 *Repeat step 2, until the current cycle cover is a Hamiltonian cycle.*

Algorithm 2 (GKS)

1 *Solve the AP with a minimum cycle cover F.*
2 *Patch two cycles of F (the two cycles and the two arcs are chosen in such a way that the patching costs become minimum).*
3 *Repeat step 2, until the current cycle cover is a Hamiltonian cycle.*

In [2] it is shown how GKS can efficiently be implemented with updating techniques so that it is nearly as fast as KSP.

If the minimum cycle cover at hand isn't a Hamiltonian tour — note that the ATSP is solved otherwise —, Glover et al. [2] suggested to apply recursively a sequence of contraction and patching operations to the current minimum cycle cover with purpose to transform the minimum cycle cover into a Hamiltonian cycle. Their algorithm is as follows.

Algorithm 3 (COP/KSP and COP/GKS)

1 *Choose a threshold t.*
2 *Solve the AP with a minimum cycle cover F.*
3 *If there is a cycle with length at most t (a short cycle), delete a heaviest arc in every short cycle, contract the obtained paths (only in the short cycles), and go to step 2.*
4 *Apply KSP or GKS to transform F into a Hamiltonian cycle H.*
5 *Expand (recontract) the arcs of F contracted on step 3.*

Swapping the steps 4 and 5, leads to the following modifications of COP/KSP and COP/GKS (see Gutin and Zverovich [9]).

Algorithm 4 (KSP/COP and GKS/COP)

1 *Choose a threshold t.*
2 *Solve the AP with a minimum cycle cover F.*
3 *If there is a cycle with length at most t (a short cycle), delete a heaviest arc in every short cycle, contract the obtained paths (only in the short cycles), and go to step 2.*

4 *Expand (recontract) the arcs of F contracted on step 3.*
5 *Apply KSP or GKS to transform F into a Hamiltonian cycle H.*

Based on the results of computational experiments, Gutin and Zverovich [9] show that KSP/COP and GKS/COP give considerably better results than COP/KSP and COP/GKS because the recontraction operations in KSP/COP and GKS/ COP create more long cycles to be patched compared to the cycles of COP/ KSP and COP/GKS. Gutin and Zverovich [9] proposed that the KSP/COP algorithm with the threshold value of two (in their notation three) can serve as a good universal tour construction heuristic for the ATSP.

3   Tolerance Based Contract-or-Patch Heuristic

The concept of our version of the Contract-or-Patch heuristic is based on the *tolerance problem* for the AP as a relaxation of the ATSP. The tolerance problem for the AP with an optimal solution π_0 is the problem of finding for each arc $e = (i, j) \in E$ the maximum decrease $l(e)$ and the maximum increase $u(e)$ of the arc cost $c(e)$ preserving the optimality of π_0 under the assumption that the costs of all other arcs remain unchanged. The values of $l(e)$ and $u(e)$ are called the *lower* and *upper tolerances*, respectively, of an arc e with respect to the optimal solution π_0 and the function c of arc costs. Goldengorin et al. [4,5] showed that the upper and lower tolerance do not depend on the chosen optimal AP solution. Furthermore all upper tolerances (see Turkensteen et al. [18]) as well as all upper and lower tolerances (see Volgenant [19]) of an optimal AP solution can be computed in $O(n^3)$ time.

With purpose to decrease the computational complexity of upper tolerances for an optimal AP solution from $O(n^3)$ time to $O(n^2)$ time, we have decided to use the following approximation of the upper tolerances for the AP (for details and motivation see [3]):

Let $c[i, j_1(i)]$ and $c[i, j_2(i)]$ be the smallest and the second-smallest entries in row i of the costs matrix $||c(i, j)||$ representing the weights of arcs in the original graph G. The (approximated) upper tolerance of an arc (i, j) in the optimal AP solution is defined as $c[i, j_2(i)] - c[i, j]$, if $j = j_1(i)$, and $c[i, j_1(i)] - c[i, j]$ (a negative value!), otherwise.

Now, our version of the tolerance based Contract-or-Patch heuristic looks as follows:

Algorithm 5 (GKS/TBCOP)

1 *Choose a threshold t.*
2 *Solve the AP with minimum cycle cover F.*
3 *If there is a cycle with length at most t (a short cycle), contract one arc (only in the short cycles) with the largest approximated upper tolerance, and go to step 2.*
4 *Expand the arcs of F contracted on step 3.*
5 *Apply* **GKS** *to transform F into a Hamiltonian cycle H.*

The main distinction of GKS/TBCOP with respect to all COP versions is that after each solution of the AP the GKS/TBCOP contracts only a single arc instead of all paths, e.g., in the KSP/COP. This distinction leads to the relatively larger execution times, but on average returns Hamiltonian tours with essentially better quality.

Furthermore, we use the following observation in our implementation. If the contracted arc is an arc of a cycle of length larger than two, then the optimal AP solutions found before and after contraction are the same. Hence, in this case it is not necessary to resolve the AP again. This observation leads to very small CPU times for asymmetric instances (see Section 4), because in this case the length of most cycles in an optimal AP solution is larger than two.

4   Computational Experiments

The algorithms were implemented in C under Linux and tested on an GenuineIntel Intel® Xeon™ 3.2GHz machine with 4 GB of RAM.

We have used the implementation of the currently best heuristic KSP/COP of Gutin and Zverovich [9] and our implementation of the GKS/TBCOP heuristic. In both implementations the code of Jonker and Volgenant [13] for solving the AP is incorporated. Both heuristics have been tested on the following 9 classes of instances with the value of threshold varied from 2 to 7. The first seven classes of instances are exactly the classes from [2], class 8 is the class of GYZ instances introduced in [8] for which the domination number of the greedy algorithm for the ATSP is 1 (see Theorem 2.1 in [8] and [3]) and class 9 are 40 JGMYZZ examples from [12]. The exact description of the 9 classes is as follows.

1 *All* asymmetric instances from TSPLIB [17] (26 instances).
2 *All* symmetric instances from TSPLIB [17] with a dimension smaller than 3000 (99 instances).
3 160 asymmetric instances with $c(i,j)$ randomly and uniformly chosen from $\{0, 1, \ldots, 100000\}$ for $i \neq j$.
4 160 asymmetric instances with $c(i,j)$ randomly and uniformly chosen from $\{0, 1, \ldots, i \cdot j\}$ for $i \neq j$.
5 160 symmetric instances with $c(i,j)$ randomly and uniformly chosen from $\{0, 1, \ldots, 100000\}$ for $i < j$.
6 160 symmetric instances with $c(i,j)$ randomly and uniformly chosen from $\{0, 1, \ldots, i \cdot j\}$ for $i < j$.
7 160 sloped plane instances with given x_i, x_j, y_i, y_j randomly and uniformly chosen from $\{0, 1, \ldots, i \cdot j\}$ for $i \neq j$ and

$$c(i,j) = \sqrt{(x_i - x_j)^2 + (y_i - y_j)^2} - \max\{0, y_i - y_j\} + 2 \cdot \max\{0, y_j - y_i\}$$

for $i \neq j$.

Table 1. Average excess over optimum, AP, or HK, and average time

	Cl. 1 (26)		Cl. 2 (99)		Cl. 3 (160)		Cl. 4 (160)	
	Opt. %	Time sec.	Opt. %	Time sec.	AP %	Time sec.	AP %	Time sec.
KSP	4.25	0.01	61.45	0.08	4.31	0.88	2.27	1
KSP/COP(2)	4.38	0.01	66.67	0.14	3.22	0.93	1.65	1.05
KSP/COP(3)	9.33	0.01	71.64	0.11	2.73	0.96	1.38	1.09
KSP/COP(4)	6.71	0.02	74.04	0.1	2.42	1.01	1.25	1.09
KSP/COP(5)	8.58	0.02	77.6	0.1	2.21	1.01	1.2	1.26
KSP/COP(6)	9.73	0.02	77.64	0.09	2.06	1.03	1.11	1.18
KSP/COP(7)	9.03	0.02	78.47	0.09	2.02	0.99	1.02	1.1
GKS	3.36	0.01	61.02	1.29	4.23	0.98	2.23	1.09
GKS/TBCOP(2)	3.24	0.02	57.45	10.03	3.18	1.05	1.65	1.19
GKS/TBCOP(3)	3.24	0.02	54.47	10.3	2.58	1.11	1.38	1.3
GKS/TBCOP(4)	7.61	0.02	55.74	10.02	2.27	1.13	1.25	1.33
GKS/TBCOP(5)	8.26	0.02	57.74	10.57	2.08	1.19	1.2	1.44
GKS/TBCOP(6)	8.75	0.02	64.92	8.95	1.9	1.14	1.11	1.4
GKS/TBCOP(7)	8.57	0.02	66.16	8.71	1.83	1.14	1	1.43

	Cl. 5 (160)		Cl. 6 (160)		Cl. 7 (160)		Cl. 8 (200)		Cl. 9 (288)	
	HK %	Time sec.	HK %	Time sec.	AP %	Time sec.	AP %	Time sec.	AP/HK %	Time sec.
KSP	460.95	0.91	379.54	1.01	42.37	1.46	0	0.25	30.62	0.3
KSP/COP(2)	43.16	0.93	38.89	1.03	50.42	2.61	0	0.25	12.04	0.64
KSP/COP(3)	43.23	0.96	38.88	1.08	57.04	1.95	0.04	0.96	13.51	0.65
KSP/COP(4)	43.2	0.95	38.93	1.03	59.3	1.85	0.04	1.25	14.57	0.64
KSP/COP(5)	43.51	0.93	39.05	1.04	60.56	1.79	0.07	1.27	15.71	0.65
KSP/COP(6)	43.62	0.96	39.17	1.03	61.92	1.74	0.06	1.5	16.67	0.65
KSP/COP(7)	44.05	0.93	39.59	0.98	63.44	1.62	0.09	1.47	17.63	0.63
GKS	362.21	7.83	139.98	7.41	46.78	7.72	0	0.3	26.35	1.43
GKS/TBCOP(2)	12.51	69.46	12.76	123.73	42.5	543.88	0	0.3	7.44	33.09
GKS/TBCOP(3)	12.29	71.06	12.79	127.07	49.25	473.91	0	16.4	7.63	31.02
GKS/TBCOP(4)	12.44	69.21	12.83	123.88	51.34	432.79	0	14.66	8.31	30.81
GKS/TBCOP(5)	12.32	75.53	12.92	132.73	53.11	436.37	0	14.36	8.95	30.39
GKS/TBCOP(6)	12.17	68.43	13.07	118.87	54.05	397.26	0	14.26	9.44	30.02
GKS/TBCOP(7)	12.13	67.48	13.23	118.94	55.16	383.49	0	14.17	9.7	29.44

8 GYZ instances (see Theorem 2.1 in [8])

$$c(i,j) = \begin{cases} n^3, & \text{for } i = n, j = 1; \\ i \cdot n, & \text{for } j = i+1, i = 1, 2, \ldots, n-1; \\ n^2 - 1, & \text{for } i = 3, 4, \ldots, n-1; j = 1; \\ n \cdot \min\{i, j\} + 1, & \text{otherwise} \end{cases}$$

for each dimension $5, 10, \ldots, 1000$ (200 instances).

Table 2. Excess for all asymmetric TSPLIB instances (class 1), part 1

	br17	p43	ry48p	ft53	ft70	ftv33	ftv35	ftv38	ftv44
Dim.	17	43	48	53	70	34	36	39	45
KSP	0	0.11	6.99	12.99	1.88	13.14	1.56	1.5	7.69
KSP/COP(2)	5.13	0.18	9.62	13.05	2.1	7.31	1.36	1.31	8.18
KSP/COP(3)	112.82	0.75	11.93	15.73	2.1	10.19	1.36	1.31	11.53
KSP/COP(4)	0	3.24	14.99	15.73	1.9	17.03	9.3	16.41	10.66
KSP/COP(5)	0	3.24	14.99	14.03	2.45	23.79	16.84	19.28	10.66
KSP/COP(6)	0	0.66	14.99	14.03	3.17	23.79	19.69	19.28	15.5
KSP/COP(7)	0	0.66	14.99	14.03	3.17	23.79	19.69	15.69	15.5
GKS	0	0.32	4.52	12.31	2.84	8.09	1.09	1.05	5.33
GKS/TBCOP(2)	0	0.32	2.93	14.34	2.48	8.09	1.15	1.05	6.7
GKS/TBCOP(3)	0	0.62	2.93	12.98	1.76	8.16	1.36	1.05	6.7
GKS/TBCOP(4)	107.69	2.33	3.75	12.98	1.51	6.77	4.89	6.21	3.22
GKS/TBCOP(5)	107.69	2.33	6.08	12.61	1.07	18.58	10.52	6.21	3.22
GKS/TBCOP(6)	107.69	0.2	6.08	12.61	1.07	18.58	16.63	6.21	3.22
GKS/TBCOP(7)	107.69	0.2	6.08	12.61	1.24	18.9	18.47	6.21	3.22

	ftv47	ftv55	ftv64	ftv70	ftv100	ftv110	ftv120	ftv130	ftv140
Dim.	48	56	65	71	101	111	121	131	141
KSP	3.04	3.05	3.81	3.33	3.52	5.41	7.62	5.46	4.46
KSP/COP(2)	4.9	3.86	3.32	1.38	6.21	6.38	8.36	5.42	5.29
KSP/COP(3)	4.9	8.15	1.96	5.69	7.72	6.59	8.54	5.59	5.99
KSP/COP(4)	8.67	7.34	4.68	5.69	7.72	6.89	8.54	5.59	5.99
KSP/COP(5)	9.91	11.94	4.51	11.79	11.69	6.54	9.7	10.23	6.49
KSP/COP(6)	9.91	11.94	5.98	18.82	23.88	6.54	9.7	11.31	6.49
KSP/COP(7)	9.91	11.94	8.32	18.82	6.43	6.54	10.3	11.31	6.49
GKS	1.69	3.05	2.61	2.87	5.31	5.67	5.12	4.9	4.67
GKS/TBCOP(2)	4.05	0.19	0.76	0.92	4.14	4.55	5.36	4.03	3.88
GKS/TBCOP(3)	4.05	0.06	0.76	0.92	4.14	4.55	8.49	4.03	3.93
GKS/TBCOP(4)	5.57	0.06	0.76	0.92	4.14	4.55	8.49	4.03	4.42
GKS/TBCOP(5)	5.57	0.06	0.82	2.67	3.75	4.19	5.12	3.73	4.13
GKS/TBCOP(6)	5.57	0.06	0.82	4	3.75	4.19	5.12	3.73	4.13
GKS/TBCOP(7)	5.57	0.06	0.82	4.92	3.75	4.19	5.12	3.73	4.13

9 JGMYZZ instances (see [12]). There are 12 problem generators from [12] called *tmat, amat, shop, disc, super, crane, coin, stilt, rtilt, rect, smat,* and *tsmat.* Each of these generators yields 24 instances, 10 of dimensions 100 and 316, 3 of dimension 1000, and 1 of dimension 3162 (288 instances).

The 160 instances of the classes 3, 4, 5, 6, and 7 are composed of 10 instances for each dimension 100, 200, ..., 1000 and 3 instances for each dimension 1100, 1200, ..., 3000, respectively.

Table 1 gives the average excess of the heuristic solutions above the optima for the TSPLIB classes 1 and 2 for which the optima are known (see [17]), the AP

Table 3. Excess for all asymmetric TSPLIB instances (class 1), part 2

	ftv150	ftv160	ftv170	kro124p	rbg323	rbg358	rbg403	rbg443
Dim.	151	161	171	100	323	358	403	443
KSP	4.75	1.71	2.4	16.11	0	0	0	0
KSP/COP(2)	4.79	2.83	2.36	10.43	0	0	0	0
KSP/COP(3)	5.86	3.54	2.36	7.26	0	0	0.61	0
KSP/COP(4)	6.17	2.46	4.17	10.65	0	0	0.61	0
KSP/COP(5)	7.12	3.39	3.92	20.1	0	0.17	0.24	0
KSP/COP(6)	7.12	3.39	3.92	20.1	0	1.38	0.97	0.51
KSP/COP(7)	7.12	3.39	3.92	20.1	0	1.38	0.7	0.51
GKS	4.33	1.49	1.38	8.69	0	0	0	0
GKS/TBCOP(2)	3.49	2.5	3.05	10.22	0	0	0	0
GKS/TBCOP(3)	3.98	2.53	3.05	8.19	0	0	0	0
GKS/TBCOP(4)	3.98	2.46	3.01	6.06	0	0	0	0
GKS/TBCOP(5)	3.72	2.98	2.76	6.92	0	0	0	0
GKS/TBCOP(6)	3.72	2.98	2.76	14.33	0	0	0	0
GKS/TBCOP(7)	3.72	2.98	2.76	6.49	0	0	0	0

lower bound for the asymmetric classes 3, 4, 7, and 8, and the HK (Held-Karp) lower bound [10] for the symmetric classes 5, 6, and 9, respectively. Additionally, it gives the average execution times for all tested instances per class.

Table 1 shows that in almost all cases we get an essential improvement in quality, if we compare the algorithms with the same thresholds — in the first column of Table 1, the heuristics are parameterized by the threshold value used. For the asymmetric classes 1, 3, 4 we reach this target in comparable runtime. For the symmetric and nearly-symmetric classes 2, 5, 6, 7, 8, 9 we receive a very large improvement in quality by spending more CPU times. The reason for the large CPU times is that the current AP solution includes a large number of cycles with length 2 and hence the GKS/TBCOP should resolve a large number of APs. As mentioned in Section 3, in most cases of the asymmetric instances the current AP solution does not include cycles with length 2 and hence the GKS/TBCOP uses a slightly updated current AP solution.

It can be seen in Table 1 that for all classes of instances except classes 3 and 4 the quality of tours returned by the KSP/COP and GKS/TBCOP deteriorates when the threshold is increased. The only cases when increasing the threshold clearly results in shorter tours are the random asymmetric classes 3 and 4. For these classes the tours produced even with small threshold values are already very close to the lower bound. This suggests that a small value of the threshold will result in the best all-round performance. Gutin and Zverovich [9] recommend for the KSP/COP that the threshold value of two is used as a good universal choice. For our GKS/TBCOP the threshold value of two is also a good universal choice. This is also demonstrated by Table 2 and 3 which give an overview of the quality of both algorithms for all 26 asymmetric TSPLIB instances (class 1).

To summarize, the GKS/TBCOP heuristic with the threshold parameter fixed to 2 appears to be a reliable tour construction heuristic for a wide variety of families of ATSP instances.

5   Conclusions and Future Research Directions

In this paper we have proposed a tolerance based version of the Contract-or-Patch heuristic, denoted by GKS/TBCOP, for finding high quality Hamiltonian cycles (tours) for the ATSP, and have shown that it offers an improvement over the currently best KSP/COP heuristic in the quality of the tours it produces. Hence, we have confirmed that the largest upper tolerance of an arc is a better indicator for the insertion of such an arc in the unknown heuristic solution compared to the deletion of an arc with the largest cost. We have performed an extensive computational evaluation of both heuristics.

The most essential improvement in quality on average at least by the factor 3 is attained on the classes of symmetric and almost-symmetric instances but the CPU times become high (on average by the factor 97). This drawback of our GKS/TBCOP heuristic could be excluded if we find an efficient update of the current Assignment Problem (AP) solution after contracting an arc from a cycle of length 2 similarly to our update after contracting an arc from a cycle of length at least 3. We have studied the influence of the COP parameter, threshold, on the quality of the tours and the execution times produced by both heuristics. The results of our evaluation have shown that the costs based COP heuristics perform very similarly to their tolerance based counterparts only for the pure asymmetric instances. Based on the results of our computational experiments, we have proposed that the GKS/TBCOP heuristic with the threshold value of two can serve as a good tour construction heuristic.

An interesting direction of research is to study the relationships between different measures of symmetry of an ATSP instance — that is, e.g., the degree of symmetry which is defined as the fraction of off-diagonal entries of the cost matrix $||c(i,j)||$ that satisfy $c(i,j) = c(j,i)$. Another direction is to consider the number of cycles of length 2 which appear in the optimal AP solution with purpose to predict the computational times of finding high quality Hamiltonian cycles by the GKS/TBCOP heuristic.

Acknowledgements

This work is done when the first and the second author have enjoyed the hospitality of Applied Mathematics Department, Khmelnitsky National University, Ukraine. We would like to thank all colleagues from this department including V. G. Kamburg, S. S. Kovalchuk, and I. V. Samigulin. The research of all authors was supported by a DFG grant SI 657/5, Germany and SOM Research Institute, University of Groningen, the Netherlands.

References

1. J. Bang-Jensen, G. Gutin. Digraphs: Theory, Algorithms and Applications. Springer, London, 2002.

2. F. Glover, G. Gutin, A. Yeo, A. Zverovich. Construction heuristics for the asymmetric TSP. European J. Oper. Res. **129**, 555–568, 2001.

3. B. Goldengorin, G. Jäger, How To Make a Greedy Heuristic for the Asymmetric Traveling Salesman Competitive, SOM Research Report 05A11, University of Groningen, Groningen, The Netherlands, 2005

    (http://som.eldoc.ub.rug.nl/reports/themeA/2005/05A11/05A11.pdf).

4. B. Goldengorin, G. Jäger, P. Molitor. Some Basics on Tolerances. The Second International Conference on Algorithmic Aspects in Information and Management, AAIM'06, Hong Kong, China, June 20-22, 2006. S.-W. Cheng, C.K. Poon (Eds.), Lecture Notes in Comput. Sci. **4041**, 194–206, 2006.

5. B. Goldengorin, G. Jäger, P. Molitor. Tolerances Applied in Combinatorial Optimization. J. Comput. Sci. 2 (9), 716–734, 2006.

6. B. Goldengorin, G. Sierksma, M. Turkensteen. Tolerance Based Algorithms for the ATSP. Graph-Theoretic Concepts in Computer Science. 30th International Workshop, WG 2004, Bad Honnef, Germany, June 21-23, 2004. J. Hromkovic, M. Nagl., B. Westfechtel (Eds.), Lecture Notes in Comput. Sci. **3353**, 222–234, 2004.

7. H.J. Greenberg, An annotated bibliography for post-solution analysis in mixed integer and combinatorial optimization. In: D. L. Woodruff (ed.), Advances in Computational and Stochastic Optimization, Logic Programming, and Heuristic Search. Kluwer Academic Publishers, 97–148, 1998.

8. G. Gutin, A. Yeo, A. Zverovich. Traveling salesman should not be greedy: domination analysis of greedy type heuristics for the TSP. Discrete Appl. Math. **117**, 81–86, 2002.

9. G. Gutin, A. Zverovich. Evaluation of the contract-or-patch heuristic for the Asymmetric TSP. INFOR **43**(1), 23–31, 2005.

10. M. Held, R. Karp. The Traveling-Salesman Problem and Minimum Spanning Trees. Oper. Res. **18**, 1138–1162, 1970.

11. K. Helsgaun. An effective implementation of the Lin-Kernighan traveling salesman heuristic. European J. Oper. Res. **126**, 106–130, 2000.

12. D.S. Johnson, G. Gutin, L.A. McGeoch, A. Yeo, W. Zhang, A. Zverovich. Experimental analysis of heuristics for the ATSP. Chapter 10 in: The Traveling Salesman Problem and Its Variations. G. Gutin, A.P. Punnen (Eds.). Kluwer, Dordrecht, 445–489, 2002.

13. R. Jonker, A. Volgenant. A Shortest Augmenting Path Algorithm for Dense and Sparse Linear Assignment Problems. Computing **38**, 325-340, 1987.

14. R.M. Karp. The probabilistic analysis of some combinatorial search algorithms. In J.F. Traub, Algorithms and Complexity: New Directions and Recent Results. Academic Press, New York, 1–19, 1976.

15. R.M. Karp. A patching algorithm for the non-symmetric traveling salesman problem. SIAM J. Comput. **8**, 561-573, 1979.

16. M. Libura. Sensitivity analysis for minimum hamiltonian path and traveling salesman problems. Discrete Appl. Math. **30**, 197–211, 1991.

17. G. Reinelt. TSPLIB – a Traveling Salesman Problem Library. ORSA J. Comput. **3**, 376–384, 1991.
18. M. Turkensteen, D. Ghosh, B. Goldengorin, G. Sierksma. Tolerance-Based Branch and Bound Algorithms. A EURO conference for young OR researches and practitioners, ORP3 2005, September 6 – 10, 2005, Valencia, Spain. Proceedings Edited by C. Maroto et al., ESMAP S.L., 171–182, 2005.
19. Volgenant, A.: An addendum on sensitivity analysis of the optimal assignment. European J. Oper. Res. 169, 338–339, 2006.

Acyclic Type-of-Relationship Problems on the Internet

Sven Kosub*, Moritz G. Maaß**, and Hanjo Täubig***

Fakultät für Informatik, Technische Universität München,
Boltzmannstraße 3, D-85748 Garching, Germany
{kosub, maass, taeubig}@in.tum.de

Abstract. We contribute to the study of inferring commercial relationships between autonomous systems (AS relationships) from observable BGP routes. We deduce several forbidden patterns of AS relationships that impose a certain type of acyclicity on the AS graph. We investigate algorithms for solving the *acyclic all-paths type-of-relationship* problem, i.e., given a set of AS paths, find an orientation of the edges according to some types of AS relationships such that the oriented AS graph is acyclic (with respect to the forbidden patterns) and all AS paths are valley-free. As possible AS relationships we include customer-to-provider, peer-to-peer, and sibling-to-sibling. Moreover, we examine a number of problem versions parameterized by sets K and U where K is the set of edge types available for describing explicit pre-knowledge and U is the set of edge types available for completion of partial orientations. A complete complexity classification of all 56 cases (8 type sets for pre-knowledge and 7 type sets for completion) is given. The most relevant practical result is a linear-time algorithm for finding an acyclic and valley-free completion using customer-to-provider relations given *any* kind of pre-knowledge. Interestingly, if we allow sibling-to-sibling relations for completions then most of the non-trivial inference problems become NP-hard.

1 Introduction

Numerous studies (e.g., [7,9,14] to name only few) have exposed that, in order to understand the dynamics of Internet inter-domain routing, it is not sufficient to possess deeper knowledge of the AS graph, i.e., the physical-connection topology among autonomous systems (ASes). Rather, most importantly, local routing policies of independent administrative domains and their interplay based on business relations have a critical influence on route (in)stability and routing quality. As business contracts are intentionally considered to be commercial secrets, many researchers have proposed methods to elicit this crucial information indirectly (see, e.g., [2,3,4,16,12]).

* Supported by Minerva Foundation of Max-Planck-Gesellschaft. Work done in part while visiting The Hebrew University of Jerusalem.
** Supported by DFG, grant Ma 870/5-1 (Leibnizpreis Ernst W. Mayr).
*** Supported by DFG, grant Ma 870/5-1 (Leibnizpreis Ernst W. Mayr) and by DFG, grant Ma 870/6-1 (SPP 1126 Algorithmik großer und komplexer Netzwerke).

T. Erlebach (Ed.): CAAN 2006, LNCS 4235, pp. 98–111, 2006.
© Springer-Verlag Berlin Heidelberg 2006

In [4], the seminal work on inferring contractual relationships on basis of sets of AS paths observable from BGP updates, heuristic approaches were devised to classify relationships into customer-to-provider, peer-to-peer, and sibling-to-sibling. A key observation is that, under rational economic behavior, AS paths exhibit a regular, so-called valley-free structure, i.e., after traversing a provider-to-customer or peer-to-peer edge, the AS path cannot traverse a customer-to-provider or peer-to-peer edge. Thus, each path has some top provider. To identify top providers in AS paths, heuristics follow basically two evident assumptions (see, e.g., [4,6,13]):

A Providers are much larger than customers in terms of infrastructure.
B The size of an AS is proportional to its degree in the AS graph.

In principle the algorithms in [4] iteratively search for vertices with maximum degree in a given AS path set to identify top providers. This approach has been further elaborated in [12,16]. Degree-based heuristics are certainly practicable if we are able to get a representative sample of all AS paths of the Internet (see [1] for a critical discussion). But, due to their over-sensitivity to path sets, they have weaknesses in well-defined analytical test situations.

A purely combinatorial treatment, neither involving **A** nor **B**, is done in [2]. There, the authors describe a linear-time algorithm solving the *all-paths type-of-relationship* problem, suggested in [12]: given a path set P, is there an orientation of the edges (indicating provider-to-customer or customer-to-provider relation) such that all paths of P are valley-free. The algorithm is based on a reduction to 2-SAT which is well-known to be solvable in linear time. In contrast, finding an orientation maximizing the number of valley-free paths is not polynomial-time approximable within factor $O(n^{1-\varepsilon})$ (unless NP = ZPP). Positively, for path lengths at most ℓ, there is a polynomial-time algorithm with approximation ratio $\frac{\ell+1}{2\ell}$ (sometimes better via SDP for MAX 2-SAT). Though computationally elegant and robust, these algorithms often lead to unrealistic relationships (i.e., well-known global providers appearing as customers of small ASes [3]).

To get closer to reality, several proposals have been made. One of them calls for partialness-to-entireness algorithms [16,2]. The basic idea is to infer the entire AS relationships from partial information obtainable from data sources other than BGP paths (see, e.g., [11,16]). Another proposal links degree-based heuristics and combinatorial optimization by considering weighted MAX 2-SAT with weights depending on degree gradients [3]. Positive experimental results suggest the fruitfulness of incorporating degree information into a combinatorial setting. Nevertheless, over-sensitivity to path sets when using assumption **B** is still problematic.

Our contribution also lies in the middle of the heuristic and the combinatorial approaches to the inference problem. Instead of using both assumptions **A** and **B**, we will only employ assumption **A** to avoid path-set over-sensitivity. The size rule **A** (and some similar one's) bears enough information to impose a global structure on oriented AS graphs. In Section 3, we deduce several unrealistic relationship patterns which we will forbid to appear in AS graph orientations. A fundamental pattern of this type is an oriented cycle within the graph, as

Fig. 1. All valley-free orientations for path set $\{(1,2,3),(2,3,4),(3,4,1),(4,1,2)\}$

this would imply that some provider is its own customer. Acyclic orientations are constructed by the degree-based heuristic approach of [4,12]. In contrast, the 2-SAT-based algorithm generally constructs cycles due to the internal computation of strongly connected components. As an example, Figure 1 shows a path set for which all valley-free orientations (restricted to customer-to-provider relationships) impose a cycle on the AS graph. However, as a prerequisite for studies related to the Internet topology (see, e.g., [5]) acyclicity is a requirement for realistic AS relationships.

In this paper, we focus on algorithms solving the *acyclic all-paths type-of-relationship* problem (ACYCLIC ToR), i.e., given a path set, find an orientation of the edges according to some types of AS relationships such that the oriented AS graph is acyclic (with respect to our forbidden patterns) and all AS paths are valley-free. As possible AS relationships we include customer-to-provider, peer-to-peer, and sibling-to-sibling. Furthermore, we meet the partialness-to-entireness requirement. We examine a number of versions ACYCLIC ToR (K, U) where K is the set of edge types available for describing explicit pre-knowledge and U is the set of edge types available for completion of partial orientations. In Section 4, we give a complete complexity classification of all 56 cases (8 type sets for pre-knowledge and 7 type sets for completion). The most relevant practical result is a linear-time algorithm for finding an acyclic and valley-free completion using customer-to-provider relations given *any* kind of pre-knowledge. Interestingly, if we allow sibling-to-sibling relations for completions, then most of the non-trivial inference problems become NP-hard.

2   An Abstract Model of BGP

We briefly describe a simple, abstract model of inter-domain routing in the Internet using BGP (see, e.g., [15,7,4]).

2.1   The Selective Export Rule

The elementary entities in our Internet world are IP addresses, i.e., bit strings of prescribed length. An autonomous system (AS) is a connected group of one or more IP prefixes (i.e., blocks of contiguous IP addresses) run by one or more network operators which has a single and clearly defined routing policy. An AS aims at providing global reachability for its IP addresses. To achieve this goal, ASes having common physical connections exchange routing informations as governed by their own local routing policies. BGP is the *de facto* standard

AS v exports to	provider	customer	peer	sibling
own routes	Yes	Yes	Yes	Yes
customer routes	Yes	Yes	Yes	Yes
provider routes	No	Yes	No	Yes
peer routes	No	Yes	No	Yes

Fig. 2. The Selective Export Rule

protocol to manage data traffic between ASes for inter-domain routing as well as for route propagation.

Reachability in the Internet depends on (physical) connectivity and contractual relationships between ASes. The most fundamental binary business relationships are customer-to-provider—where the provider sells routes to the customer—, peer-to-peer—where the involved ASes provide special routes to their customers but no transit for each other—, and sibling-to-sibling—where both ASes belong to the same administrative domain. Evidently, sibling-to-sibling relations are transitive. More peculiar types of relationships appear in the real world (see, e.g., [4]). We restrict ourselves to the three mentioned types.

More specifically, let V be a set of AS numbers. For any $v \in V$, let $N(v) \subseteq V$ denote the set of its neighbor ASes, i.e., all numbers of ASes sharing a physical connection with v. The undirected graph $G = (V, E)$ where $E = \{ \{u, v\} \mid v \in N(u) \}$ is called a *connectivity graph* (at the AS level) or simply *AS graph*. Let $v \in V$ be any AS. According to the business relationship we divide the neighbors of v into the sets $\mathrm{Cust}(v)$ of all customers of v, $\mathrm{Prov}(v)$ of all providers of v, $\mathrm{Sibl}(v)$ of all siblings of v, and $\mathrm{Peer}(v)$ of all peering partners of v. Some of the sets may be empty. We let $\mathrm{Sibl}(v)$ contain v as well. Let $R(v)$ denote the set of all currently *active* AS paths in the BGP routing table of v, i.e., all AS paths that have been announced from neighboring ASes at a certain time and never been withdrawn. Assumed that there are no misconfigurations of BGP, all AS paths in $R(v)$ are loopless and not including v. Here, we say that an AS path is *loopless* whenever between two sibling ASes on the path, no non-sibling AS is passed. Based on the neighborhood classification, we further divide $R(v)$ into four categories. A loopless AS path $(u_1, \ldots, u_r) \in R(v)$ is

$$
\begin{aligned}
&\text{a } \textit{customer route} \text{ of } v \Longleftrightarrow_{\mathrm{def}} \quad \text{leftmost } u_i \notin \mathrm{Sibl}(v) \text{ lies in } \mathrm{Cust}(v), \\
&\text{a } \textit{provider route} \text{ of } v \Longleftrightarrow_{\mathrm{def}} \quad \text{leftmost } u_i \notin \mathrm{Sibl}(v) \text{ lies in } \mathrm{Prov}(v), \\
&\text{a } \textit{peer route} \text{ of } v \Longleftrightarrow_{\mathrm{def}} \quad \text{leftmost } u_i \notin \mathrm{Sibl}(v) \text{ lies in } \mathrm{Peer}(v), \\
&\text{an } \textit{own route} \text{ of } v \Longleftrightarrow_{\mathrm{def}} \quad \text{for all } 1 \leq i \leq r, \ u_i \in \mathrm{Sibl}(v).
\end{aligned}
$$

Now, typically (at least, recommendably), ASes set up their export policies according to the Selective Export Rule (see, e.g., [4]) as described in Figure 2. In our simplified model, the receiving AS gets from an AS those (locally preferred) routes destined for it, prolongated with the number of the sending AS as the new leftmost AS number in the path.

2.2   The Valley-Free Path Model

Valley-freeness is a graph-theoretical consequence of the Selective Export Rule. Let $G = (V, E)$ be an undirected (simple) graph. We assume that $(u, v) \in E \Leftrightarrow (v, u) \in E$. A (mixed) *orientation* φ of G is a mapping from E to T where T denotes the set of possible edge-types. For instance, a directed graph is a graph oriented with type set $T = \{\leftarrow, \rightarrow\}$. We consider type sets having the following edge-types and interpretations:

> $\rightarrow$ indicating a customer-to-provider relationship
> $\leftarrow$ indicating a provider-to-customer relationship
> $-$ indicating a peer-to-peer relationship
> $\leftrightarrow$ indicating a sibling-to-sibling relationship

Throughout this paper, we only consider orientations φ that are consistent with respect to $\rightarrow$. That is, for all $(u, v) \in E$ if $\varphi(u, v) = \leftarrow$ then $\varphi(v, u) = \rightarrow$ and if $\varphi(u, v) = \rightarrow$ then $\varphi(v, u) = \leftarrow$. Thus, if we allow $\rightarrow$ as a possible edge type, then we immediately allow $\leftarrow$ as a possible edge type as well.

We extend φ from edges to walks homomorphically. Let $(v_0, v_1, \ldots, v_m)$ be any walk in a graph G. Then $\varphi(v_0, v_1, \ldots, v_m)$ is defined to be the sequence $\varphi(v_0, v_1)\varphi(v_1, v_2) \ldots \varphi(v_{m-1}, v_m)$, i.e., generally a word in $\{\leftarrow, \rightarrow, -, \leftrightarrow\}^*$. We will typically use regular expressions to describe walk types given an orientation. An important property of orientations is valley-freeness, which we state here in terms of regular patterns of paths.

Definition 1. [4] *Let G be any graph, and let $\varphi(G)$ be an orientation of G. A loopless path $(v_0, \ldots, v_m)$ is said to be* valley-free *in $\varphi(G)$ if and only if $\varphi(v_0, \ldots, v_m)$ belongs to*

$$\{\rightarrow, \leftrightarrow\}^* \{\leftarrow, \leftrightarrow\}^* \ \cup \ \{\rightarrow, \leftrightarrow\}^* - \{\leftarrow, \leftrightarrow\}^*.$$

The valley-freeness of paths abstracts the condition that autonomous systems never route data from one of their providers to another of their providers because instead of earning money, they would have to pay twice for these data streams.

Theorem 2. [4] *Let $G = (V, E)$ be an AS graph. Let P be any subset of AS paths of all BGP routing tables, i.e., $P \subseteq \bigcup_{v \in V} R(v)$. If all ASes export their routes according to the Selective Export Rule, then there exists an orientation of P such that all AS paths in P are valley-free.*

3   Acyclicity Conditions

In the previous section, we have seen how some rational economic behavior implies valley-freeness of locally observable routes in our simple, abstract BGP. These routes reflect short-term behavior determined by routing policies based on commercial relationships. Commercial relationships typically are stable over a longer period and they impose a global structure on the connectivity graph independent of concrete BGP routes.

	Typically, C is provider of B only if C is much larger than B, A is provider of C only if A is much larger than C. So, B is not much larger than A. A contradiction to B being provider of A.	$\to^*$
	A typical criterion for a peer-to-peer relation is roughly the same size or traffic. This does not hold if A is much larger than C and B is much larger than A.	$\{\to, -\}^* \to \{\to, -\}^*$
	Typically, as B and C are siblings, they behave like one AS. So, B and C together are not much larger and much smaller than A at the same time.	$\{\to, \leftrightarrow\}^* \to \{\to, \leftrightarrow\}^*$
	Typically, C and A have a peer-to-peer relation if they are roughly the same size. The same holds for A and B. So, B should not be much larger than C.	$\{\to, -\}^* \to \{\to, -\}^*$
	Typically, A and B have a peer-to-peer relation if A and B together with its sibling C have roughly the same size. So, A is not much larger than C together with its sibling B.	$\{\to, \leftrightarrow, -\}^* \to \{\to, \leftrightarrow, -\}^*$
	Typically, A and C have a peer-to-peer relation if A and C together with its sibling B have roughly the same size. So, A is not much smaller than B together with its sibling C.	$\{\to, \leftrightarrow, -\}^* \to \{\to, \leftrightarrow, -\}^*$
	Due to the transitivity of the sibling-to-sibling relation, A and B are siblings. So, typically, A and B do not have a peculiar peer-to-peer relation. Note that larger cycles with siblings and at least two peer-to-peer relations make sense.	$\leftrightarrow^* - \leftrightarrow^*$
	Due to the transitivity of the sibling-to-sibling relation, A and B are siblings. So, A together with its siblings B and C is not much larger than C together with its siblings B and A.	$\{\to, \leftrightarrow\}^* \to \{\to, \leftrightarrow\}^*$

Fig. 3. Forbidden triads and their generalized forbidden patterns

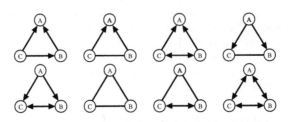

Fig. 4. Allowed triads

In this section, we summarize common knowledge on business relations between ASes to obtain a reasonable acyclicity structure within a connectivity graph. We do so by identifying patterns of oriented cycles which we will forbid to be contained in the graph. An oriented cycle can be interpreted as someone being its own provider and customer. In Figures 3 and 4, the 16 non-isomorphic triads of the 64 possible orientations of a complete graph on three vertices are shown. Figure 3 lists 8 forbidden triads together with plausibility arguments why they are forbidden. The generalizations of the forbidden patterns are fairly obvious. Plausibility is based on size rules:

1. If AS u is a customer of AS v, then AS u has much smaller size (i.e., number of routers) than AS v (see, e.g., [4,6,12]). This is assumption **A** from the introductory section.
2. If AS u is a peering partner of AS v, then AS u and AS v are roughly of the same size (see, e.g., [10]). Moreover, we consider *roughly the same size* to be a transitive relation.
3. If AS u and AS v are siblings, then they count as one AS, i.e., we assume that the size of u is determined by its own number of routers and the number of routers of all its sibling ASes.

Each of these rules may have its exceptions but they certainly describe typical behavior. We idealize the BGP world by assuming that all contracts follow these rules. The union of all generalized forbidden patterns given in Figure 3 leads to the following definition of an oriented cycle.

Definition 3. *Let G be any graph, and let $\varphi(G)$ be an orientation of G. Let C be any minimal cycle of G, i.e., a cycle that does not contain a vertex twice. C is said to be an* oriented cycle *of $\varphi(G)$ if and only if $\varphi(C)$ belongs to*

$$\{-, \leftrightarrow\}^* \to \{\to, -, \leftrightarrow\}^* \quad \cup \quad \{-, \leftrightarrow\}^* \leftarrow \{\leftarrow, -, \leftrightarrow\}^* \quad \cup \quad \leftrightarrow^* - \leftrightarrow^*.$$

The minimality of cycles is required since we exactly count occurrences of peer-to-peer edges in oriented cycles. To complement our view on forbidden triads, Figure 4 shows the 8 allowed triads.

Note that in the case that φ does not exhaust the full type set $\{\to, -, \leftrightarrow\}$, the patterns of oriented cycles simplify. For instance, if the type set is $\{\to\}$, then we obtain that a minimal cycle C is an oriented cycle if and only if $\varphi(C)$ belongs to $\to^*$ or $\leftarrow^*$ which is the usual understanding of a cycle. As a second example, if the type set is $\{\leftrightarrow, -\}$, then a minimal cycle C is an oriented cycle if and only if $\varphi(C)$ belongs to $\leftrightarrow^* - \leftrightarrow^*$.

We call an orientation *acyclic* if it contains no oriented cycles. In the forthcoming we will need fast algorithms for testing acyclicity which are all based on standard techniques.

Lemma 4. *Let K be any subset of $\{\to, -, \leftrightarrow\}$. Testing whether a given graph, with n vertices and m edges, which is oriented with type set K is acyclic can be done in time $O(n + m)$.*

4 Combinatorial Inference of Relationships

Our central problem is: given any set of BGP routes, is there an orientation such that all BGP paths are indeed valley-free and the induced connectivity graph has the additional property of being acyclic? We bring this issue into a very general, purely combinatorial formulation. Let K and U be two type sets.

Problem:	ACYCLIC TOR (K, U)
Input:	Undirected graph G, path set P, partial orientation φ given by an edge set R with labels from K
Output:	A completion of the orientation φ using only edge types from U such that the completed orientation, with respect to type set $K \cup U$, contains no cycle and all oriented paths in P are valley-free, or indicate that such a completion does not exist

Interpretation. Applied to a real-world scenario, G is the AS graph, P stands for the set of observed BGP routes, e.g., gathered at certain observation points. The set R is an explicit pre-knowledge we have of certain relations between two ASes, as observable from several resources on the Internet. The task is to find a hypothetical valley-free and acyclic orientation in accordance with our pre-knowledge. In a test setting, G is a BGP-world model, P could be the set of information made available to BGP speakers, and R could describe a specified situation with types from K. Here, we want to find an orientation that guarantees valley-freeness and acyclicity and which does not destroy the given specification. For this purpose, an appropriate choice of a type set U is useful. The most important case, of course, is $U = \{\to\}$.

Input representation. Let N always denote the input size, i.e., $N = \|V(G)\| + \|E(G)\| + |P| + |R|$ where $|P|$ and $|R|$ are the sums of the path lengths in P and in R. The length of a path with k edges is the sum of the lengths of the $k + 1$ vertex descriptions. In order to guarantee linear time-bounds for many of our algorithms, we fix the instance representation. For the sake of simplicity we assume that all vertices and edges of G actually appear in P, i.e., $G = G(P) = (V(P), E(P))$ where $V(P)$ denotes the set of vertices appearing in P and $E(P)$ denotes the set of edges appearing in P. Thus, $N = O(|P|)$. We use $V(p)$ and

$E(p)$ to denote $V(\{p\})$ and $E(\{p\})$. We suppose that an instance is given as an adjacency list of an AS graph $G(P)$ with vertex set $V(P)$ and edge set $E(P)$, where edges from R are already labeled according to the partial orientation. Moreover, we assume that the path set is represented as a collection of lists having cross-links to the corresponding edges in the AS graph and vice versa. Orientations are stored with the edges in $G(P)$. Since there are not more vertices and edges in $G(P)$ than appearing in P, this guarantees that we can always test valley-freeness in time $O(N)$.

K	$\rightarrow$	$-$	$\leftrightarrow$	$\rightarrow, -$	$\rightarrow, \leftrightarrow$	$-, \leftrightarrow$	$\rightarrow, -, \leftrightarrow$
				U			
$\emptyset$	$O(N)$	$O(N)$	$O(N)$	$O(N)$	$O(N)$	$O(N)$	$O(N)$
$\rightarrow$	$O(N)$	$O(N)$	$O(N)$	$O(N)$	NP-hard	$O(N)$	NP-hard
$-$	$O(N)$	$O(N)$	$O(N)$	$O(N)$	NP-hard	NP-hard	NP-hard
$\leftrightarrow$	$O(N)$	$O(N)$	$O(N)$	$O(N)$	$O(N)$	$O(N)$	$O(N)$
$\rightarrow, -$	$O(N)$	$O(N)$	$O(N)$	$O(N)$	NP-hard	NP-hard	NP-hard
$\rightarrow, \leftrightarrow$	$O(N)$	$O(N)$	$O(N)$	$O(N)$	NP-hard	$O(N)$	NP-hard
$-, \leftrightarrow$	$O(N)$	$O(N)$	$O(N)$	$O(N)$	NP-hard	NP-hard	NP-hard
$\rightarrow, -, \leftrightarrow$	$O(N)$	$O(N)$	$O(N)$	$O(N)$	NP-hard	NP-hard	NP-hard

Fig. 5. Complexity classification of ACYCLIC TOR (K, U) for $K, U \subseteq \{\rightarrow, -, \leftrightarrow\}$

Technical remarks. The complexity of the problem depends on the types of allowed orientations. The basic (and typically considered) combinatorial problem appears as case $K = \emptyset$, i.e., where we have no pre-knowledge. Without going into formal details, we easily see that if $K \subseteq K'$, then ACYCLIC TOR (K, U) is computationally not harder than ACYCLIC TOR (K', U), i.e., an algorithmic upper bound for ACYCLIC TOR (K', U) is an upper bound for ACYCLIC TOR (K, U) and an algorithmic lower bound for ACYCLIC TOR (K, U) is an algorithmic lower bound for ACYCLIC TOR (K', U). We say that ACYCLIC TOR (K, U) is NP-hard if it is NP-hard to decide whether for a given instance of ACYCLIC TOR (K, U), there is an acyclic and valley-free orientation in the sense of the problem definition. Evidently, if ACYCLIC TOR (K, U) is NP-hard, then for all $K' \supseteq K$, ACYCLIC TOR (K', U) is NP-hard.

Figure 5 shows a complete complexity classification of all reasonable cases for K and U. It is interesting to observe that either there is a linear-time algorithm for a problem or it is NP-hard. The classification is thus optimal. The most remarkable results are:

1. ACYCLIC TOR $(\{\rightarrow, \leftrightarrow, -\}, \{\rightarrow\})$ can be solved in time $O(N)$. This result underlines that the partialness-to-entireness methodology of [16], aside from producing more and more realistic AS relationships, is a very feasible one.
2. ACYCLIC TOR $(\{\rightarrow\}, \{\rightarrow, \leftrightarrow\})$ is NP-hard. This is surprising since there is a linear-time algorithm for ACYCLIC TOR $(\{\rightarrow\}, \{\rightarrow\})$ and $\leftrightarrow$ very often is a negligible type of relationship.

Due to page restrictions, we only prove two cases: the fundamental linear-time case $K = \emptyset$ and $U = \{\rightarrow\}$ as well as the simplest NP-hard case $K = \{\rightarrow\}$ and $U = \{\rightarrow, \leftrightarrow\}$. The proofs of all other results can be found in the full paper [8].

To obtain a linear-time algorithm for the first case $K = \emptyset$ and $U = \{\rightarrow\}$, the crucial observation is that each vertex appearing somewhere in the middle of a path has in-degree at least one, valley-freeness supposed. This allows us to employ a topological-sort approach.

Theorem 5. ACYCLIC TOR $(\emptyset, \{\rightarrow\})$ *can be solved in time $O(N)$.*

Proof. Suppose we are given a path set P. Let v be a vertex such that, for each path in P, if v lies on p then v is an endpoint of the path. If $G(P)$ can be acyclically oriented such that all paths are oriented valley-free then such a vertex v must exists: for any acyclic orientation of $G(P)$ there must exist at least one vertex u such that all edges $\{u, w\}$ are oriented as $u \rightarrow w$. Since all paths are valley-free, u cannot be in the middle of any path because that would result in an orientation containing $\leftarrow\rightarrow$. Therefore, one vertex v can be forced.

We iteratively reduce the problem by removing such vertices v for all path-ends and orienting the removed edges away from v as $v \rightarrow w$ for all neighbors w in P. (Note that P has changed.) Assume a reduced path $(v_0, \ldots, v_m)$ is oriented as $\rightarrow^* \leftarrow^*$, then adding v to any end with the edge u, v_1 or u, v_m oriented as $\rightarrow$ results in a valley-free orientation. Furthermore, if the reduced graph is acyclic then the graph with v added is also acyclic.

A precise description is given as Algorithm 1. Clearly, this algorithm can be implemented in such a way that its running time is $O(|P|)$.   □

Algorithm 1 described in Theorem 5 can be extended by additional linear-time preprocessing phases to handle non-trivial pre-knowledge. The main obstacle for designing an $O(N)$ algorithm to solve ACYCLIC TOR $(\{\rightarrow, \leftrightarrow, -\}, \{\rightarrow\})$ is the peculiar nature of peer-to-peer relations. In one direction peer-to-peer edges behave like sibling-to-sibling relations, due to their symmetry, and in another direction they behave like customer-to-provider relations, due to the forbidden cycle $\leftrightarrow^* - \leftrightarrow^*$. However, both aspects can be combined to prove that peer-to-peer relations exhibit a very regular structure, namely they induce globally a partial ordering among vertex sets and locally a total ordering of peer-to-peer pairs on each path. This structure is algorithmically exploitable to get rid of peer-to-peer relations.

We now turn to the second case $K = \{\rightarrow\}$ and $U = \{\rightarrow, \leftrightarrow\}$. To prove the NP-hardness, observe that, as sibling-to-sibling relations establish an equivalence relation in any acyclic and valley-free orientation, we obtain a partition $[B_1, \ldots, B_r]$ of V from a given orientation, i.e., a collection of non-empty subsets of V satisfying $B_i \cap B_j = \emptyset$ for $i \neq j$, and $B_1 \cup \cdots \cup B_r = V$, where the B_i are just equivalence classes according to the sibling-to-sibling relation. If the orientation is not already known to us then we search for suitable candidates for such partitions. We say that a pair $(P, [B_1, \ldots, B_r])$ is an *admissible decomposition* of P if and only if $[B_1, \ldots, B_r]$ is a partition of V such that the path set

---

Algorithm 1: Linear-time algorithm for ACYCLIC TOR $(\emptyset, \{\rightarrow\})$

Input: Undirected graph G, path set P

Output: Acyclic and valley-free orientation of the induced graph $G(P)$, if it exists, or indication that it does not exist

1 **foreach** *vertex* $v \in V(P)$ **do**
2 $\quad$ count$(v) := 0$
3 **end**
4 **foreach** $p \in P$ **do**
5 $\quad$ **foreach** *vertex* $v \in p$ **do**
6 $\quad\quad$ **if** *v is not an endnode of p* **then**
7 $\quad\quad\quad$ count$(v) :=$ count$(v) + 1$
8 $\quad\quad$ **end**
9 $\quad$ **end**
10 **end**
11 $U := \emptyset$
12 **foreach** *vertex* $v \in V(P)$ **do**
13 $\quad$ **if** count$(v) = 0$ **then**
14 $\quad\quad$ $U := U \cup \{v\}$
15 $\quad$ **end**
16 **end**
17 $V' := \emptyset$
18 **while** $U \neq \emptyset$ **do**
19 $\quad$ remove a vertex u from U
20 $\quad$ **foreach** *vertex* v *such that* $v \in V \setminus V'$ *and* $\{u, v\} \in E(P)$ **do**
21 $\quad\quad$ orient $\{u, v\}$ as $u \rightarrow v$
22 $\quad\quad$ **foreach** $p \in P$ *such that* u *and* v *are neighbors in* p **do**
23 $\quad\quad\quad$ **if** *v has a neighbor w in p on the side opposite to u such that* $w \in V \setminus V'$ **then**
24 $\quad\quad\quad\quad$ count$(v) :=$ count$(v) - 1$
25 $\quad\quad\quad$ **end**
26 $\quad\quad\quad$ **if** count$(v) = 0$ **then**
27 $\quad\quad\quad\quad$ $U := U \cup \{v\}$
28 $\quad\quad\quad$ **end**
29 $\quad\quad$ **end**
30 $\quad$ **end**
31 $\quad$ $V' := V' \cup \{u\}$
32 **end**
33 **if** $V' \neq V$ **then**
34 $\quad$ **return** *path set P fails to allow an acyclic and valley-free orientation*
35 **end**

---

obtained by replacing all vertices of the same block with a unique representative and afterwards removing all multiple occurrences of representatives in all paths, allows an acyclic and valley-free orientation without $\leftrightarrow$. Notice that this also requires looplessness of paths. We further define the standard refinement relation $\subseteq$ on partitions which allows us to order partitions. Let $[A_1, \ldots, A_s]$ and $[B_1, \ldots, B_r]$ be two partitions of the same set V. We define

$$[A_1, \ldots, A_s] \subseteq [B_1, \ldots, B_r] \iff_{\mathrm{def}} (\forall i, 1 \leq i \leq s)(\exists j, 1 \leq j \leq r)[A_i \subseteq B_j].$$

We easily observe that admissible decompositions behave monotonically with respect to refinements, i.e., if for partitions $\mathcal{A}$ and $\mathcal{B}$, $\mathcal{A} \subseteq \mathcal{B}$ and $\mathcal{A}$ is an admissible decomposition of P, then so is $\mathcal{B}$.

Theorem 6. ACYCLIC TOR $(\{\rightarrow\}, \{\rightarrow, \leftrightarrow\})$ *is* NP-*hard*.

Proof. For the proof of the NP-hardness, we use the NP-complete problem 2-IN-3-SAT, i.e., the problem where we ask, given a 3CNF H, whether there is an assignment to all variables of H such that in each clause exactly two of the three literals are true. It will be enough to reduce 2-IN-3-SAT to the decision version of ACYCLIC TOR $(\{\rightarrow\}, \{\rightarrow, \leftrightarrow\})$. Let H be any 3CNF having $m \geq 3$ clauses $C_1, \ldots, C_m$, each having exactly three different literals, and variables $x_1, \ldots, x_n$. We construct a path set P on the vertex set $\{C_1, \ldots, C_m, x_1, \ldots, x_n, \overline{x_1}, \ldots, \overline{x_n}\}$. Define the following sets of paths:

$$P_1 =_{\mathrm{def}} \{ (C_1, C_2, C_3), (C_2, C_3, C_1), (C_3, C_1, C_2) \}$$
$$\cup \{(C_1, C_j, C_2) \mid 4 \leq j \leq m\}$$

$$P_2 =_{\mathrm{def}} \{ (x_i, x_j, \overline{x_i}) \mid 1 \leq i, j \leq n \wedge i \neq j \} \cup \{ (x_1, C_i, \overline{x_1}) \mid 1 \leq i \leq m \}$$

$$P_3 =_{\mathrm{def}} \{ (l_{i1}, C_i, l_{i2}), (l_{i2}, C_i, l_{i3}), (l_{i3}, C_i, l_{i1}), (l_{i1}, \overline{l_{i2}}, l_{i3}) \mid$$
$$1 \leq i \leq m \wedge C_i = \{l_{i1}, l_{i2}, l_{i3}\} \}$$

The set P_1 guarantees that all clause vertices C_i belong to the same set, P_2 separates the literals from their negated literals, and P_3 indicates which literals will be satisfied. Now define $P =_{\mathrm{def}} P_1 \cup P_2 \cup P_3$. Note that clearly, P can be computed in time polynomial in the number of clauses and variables of the input. We will show that

$$H \in 2\text{-IN-3-SAT} \iff$$

$$(G(P), P, \{x_1 \rightarrow \overline{x_1}\}) \text{ is a solvable ACYCLIC TOR } (\{\rightarrow\}, \{\rightarrow, \leftrightarrow\}) \text{ instance.}$$

We prove both directions separately.

For ($\Rightarrow$), let $I : \{x_1, \ldots, x_n\} \rightarrow \{0, 1\}$ be an assignment to variables witnessing that $H \in 2\text{-IN-3-SAT}$. Define U to be the set

$$\{ x_i \mid 1 \leq i \leq n \wedge I(x_i) = 1 \} \cup \{ \overline{x_i} \mid 1 \leq i \leq n \wedge I(x_i) = 0 \} \cup \{ C_i \mid 1 \leq i \leq m \}.$$

Hence, $\overline{U} = \{ \overline{x_i} \mid 1 \leq i \leq n \wedge I(x_i) = 1 \} \cup \{ x_i \mid 1 \leq i \leq n \wedge I(x_i) = 0 \}$. Clearly, $x_1 \in U \Leftrightarrow \overline{x_1} \notin U$. We are done if we can show that $[U, \overline{U}]$ is an admissible decomposition of P. In the following we use U (or $\overline{U}$) to denote any representative of U (or, $\overline{U}$, respectively). We consider all path sets individually:

1. Since $C_i \in U$ for all $1 \leq i \leq m$, all paths in P_1 have the form (U, U, U) which simplifies to (U).
2. Without loss of generality, suppose $x_i \in U$ which immediately implies that $\overline{x_i} \notin U$. It follows that the paths of P_2 have the form $(U, U, \overline{U})$ or $(U, \overline{U}, \overline{U})$ which both simplify to $(U, \overline{U})$.

3. In each clause, exactly two literals are satisfied. Without loss of generality, suppose that l_{i_1} and l_{i2} are these literals for clause C_i. Then, the paths of P_3 that correspond to C_i all have forms (U, U, U), $(U, U, \overline{U})$, or $(U, \overline{U}, \overline{U})$, thus, simplify to (U) or $(U, \overline{U})$.

Consequently, after eliminating multiple occurrences of paths, P simplifies to a subset of $\{(U), (\overline{U}), (U, \overline{U}), (\overline{U}, U)\}$ which, obviously, allows acyclic and valley-free orientations. Hence, we obtain that $(G(P), P, \{x_1 \to \overline{x_1}\})$ is a solvable ACYCLIC ToR $(\{\to\}, \{\to, \leftrightarrow\})$ instance.

For $(\Leftarrow)$, we assume that $(G(P), P, \{x_1 \to \overline{x_1}\})$ is a solvable instance of ACYCLIC ToR $(\{\to\}, \{\to, \leftrightarrow\})$, i.e., there exists a decomposition $[U, \overline{U}] \subset [V]$ of P such that $x_1 \in U \Leftrightarrow \overline{x_1} \notin U$. Note that we can restrict ourselves to such 2-component decompositions because of the $\subseteq$-monotonicity of admissible decompositions. Without loss of generality, we may assume that $C_1 \in U$. Thus, $\{C_1, \ldots, C_m\} \subseteq U$ because of the definition of P_1. Furthermore, we have for all $1 \le i \le n$, $x_i \in U \Leftrightarrow \overline{x_i} \notin U$ (otherwise $U = V$ or $\overline{U} = V$ because of definition of P_2). This allows to define an assignment $I : \{x_1, \ldots, x_n\} \to \{0, 1\}$ as $I(x_i) = 1$ if $x_i \in U$ and $I(x_i) = 0$ otherwise. We have to prove that I satisfies exactly two literals in each clause. Let C_i be any clause with literals l_{i1}, l_{i2}, and l_{i3}. As there are paths in P_3 having the form (l_{ij}, C_i, l_{ik}), there is an $l_{ir} \in U$. Without loss of generality, we assume that l_{i1} is such a literal. Since there exists for each pairs of literals of C_i such a path in P_3, there exists another literal $l_{is} \in U$, $r \ne s$. Without loss of generality, l_{i2} is such a literal. Due to the path $(l_{i1}, \overline{l_{i2}}, l_{i3}) \in P_3$, we know that l_{i3} is not in U. Overall, by definition of I, for each clause exactly two literals are made true. This shows $H \in$ 2-IN-3-SAT. □

5 Conclusion

We studied algorithmic solutions for the acyclic all-paths type-of-relationship problem. In particular we designed a linear-time algorithm for finding an acyclic and valley-free completion of a partial orientation of a set of AS paths, that only uses customer-to-provider relations for completion whereas the partial orientation can be expressed with arbitrary types of standard AS relationships. Based on some evident assumptions on the size of ASes, acyclicity conditions are given in terms of forbidden graph patterns. The algorithm provides prospects for combining combinatorial methods with more non-combinatorial techniques to explore the solution space of possible Internet hierarchies. To evaluate the quality of this algorithm, we plan to supplement the theoretical study of this paper with experimental investigations. In contrast, permitting sibling-to-sibling relations for completion makes most of the problem versions NP-hard.

Several algorithmic problems remain open. However, from a theoretical point of view, the most interesting open question is whether acyclicity of the Internet hierarchy can be deduced via a game-theoretic analysis. Are acyclically oriented AS graphs Nash equilibria for classes of network creation games?

Acknowledgment. For helpful hints and discussions we thank Vinay Aggarwal, Stefan Eckhardt, Klaus Holzapfel, Olaf Maennel, Johannes Nowak, Thomas

Schwabe, Arne Wichmann. In particular we are deeply grateful to Benjamin Hummel for carefully proofreading earlier drafts of this paper.

References

1. H. Chang, R. Govindan, S. Jamin, S. J. Shenker, W. Willinger. Towards capturing representative AS-level Internet topologies. *Comput. Netw.*, 44(6):737–755, 2004.
2. G. Di Battista, T. Erlebach, A. Hall, M. Patrignani, M. Pizzonia, T. Schank. Computing the types of the relationships between autonomous systems. *IEEE/ACM Trans. Networking*. To appear.
3. X. Dimitriopoulos, D. V. Krioukov, B. Huffaker, K. C. Claffy, G. F. Riley. Inferring AS relationships: Dead end or lively beginning? In *Proc. 4th International Workshop on Experimental and Efficient Algorithms (WEA'05)*, LNCS 3503, pp. 113–125. Springer, 2005.
4. L. Gao. On inferring autonomous system relationships in the Internet. *IEEE/ACM Trans. Networking*, 9(6):733–745, 2001.
5. L. Gao, J. Rexford. Stable Internet routing without global coordination. *IEEE/ACM Trans. Networking*, 9(6):681–692, 2001.
6. R. Govindan, A. Reddy. An analysis of Internet inter-domain topology and route stability. In *Proc. 16th Joint Conference of the IEEE Computer and Communications Societies (INFOCOM'97)*, pp. 850–857. IEEE, 1997.
7. T. G. Griffin, G. T. Wilfong. An analysis of BGP convergence properties. *Comput. Commun. Rev.*, 29(4):277–288, 1999.
8. S. Kosub, M. G. Maaß, H. Täubig. Acyclic type-of-relationship problems on the Internet. Technical Report TUM-I0605, Fakultät für Informatik, Technische Universität München, March 2006.
9. C. Labovitz, A. Ahuja, R. Wattenhofer, S. Venkatachary. The impact of Internet policy and topology on delayed routing convergence. In *Proc. 20th Joint Conference of the IEEE Computer and Communications Societies (INFOCOM'01)*, pp. 537–546. IEEE, 2001.
10. W. B. Norton. Internet Service Providers and peering. Equinix White Paper, Equinix, Inc., 2001.
11. G. Siganos, M. Faloutsos. Analyzing BGP policies: Methodology and tool. In *Proc. 23rd Joint Conference of the IEEE Computer and Communications Societies (INFOCOM'04)*, pp. 1640–1651. IEEE, 2004.
12. L. Subramanian, S. Agarwal, J. Rexford, R. H. Katz. Characterizing the Internet hierarchy from multiple vantage points. In *Proc. 21st Joint Conference of the IEEE Computer and Communications Societies (INFOCOM'02)*, pp. 618–627. IEEE, 2002.
13. H. Tangmunarunkit, J. Doyle, R. Govindan, S. Jamin, S. J. Shenker, W. Willinger. Does AS size determine degree in AS topology? *Comput. Commun. Rev.*, 31(5):7–10, 2001.
14. H. Tangmunarunkit, R. Govindan, S. Shenker, D. Estrin. The impact of routing policy on Internet paths. In *Proc. 20th Joint Conference of the IEEE Computer and Communications Societies (INFOCOM'01)*, pp. 736–742. IEEE, 2001.
15. I. van Beijnum. *BGP*. O'Reilly & Associates, 2002.
16. J. Xia, L. Gao. On the evaluation of AS relationship inferences. In *Proc. 47th IEEE Global Telecommunications Conference (Globecom'04)*, vol. 3, pp. 1373–1377. IEEE, 2004.

Minimum-Energy Broadcasting in Wireless Networks in the d-Dimensional Euclidean Space (The $\alpha \leq d$ Case)

Andrzej Lingas, Mia Persson, and Martin Wahlen

Department of Computer Science, Lund University
Box 118, 221 00 Lund, Sweden
{andrzej, mia, martin}@cs.lth.se

Abstract. We consider the problem of minimizing the total energy assigned to nodes of wireless network so that broadcasting from the source node to all other nodes is possible. This problem has been extensively studied especially under the assumption that the nodes correspond to points in the Euclidean two- or three-dimensional space and the broadcast range of a node is proportional to at most the α root of the energy assigned to the node where α is not less than the dimension d of the space. In this paper, we study the case $\alpha \leq d$, providing several tight upper and lower bounds on approximation factors of known heuristics for minimum energy broadcasting in the d-dimensional Euclidean space.

1 Introduction

Ad hoc wireless networks are the most popular type of multi-hop networks [12,15]. They do not have any wired infrastructure. A communication session is achieved either through a single hop transmission if the parties are close enough, or through relaying by intermediate nodes. Omnidirectional antennas are used by all nodes to transmit and receive signals.

In the most common power-attenuation model [19], the signal power in the distance r from the transmitter is usually proportional to $\frac{1}{r^\alpha}$, where $\alpha \geq 1$. One assumes that all receivers have the same power threshold for signal detection and consequently that the power required to establish a communication link between two nodes separated by range r is proportional to r^α. Hence, in order to send a message from one node to another, the sending node needs to emit the message with enough power such that target node can receive it.

Ad hoc wireless networks are usually powered by very limited electricity resources, e.g., batteries. Therefore low power consumption has become a crucial issue in the design of routing communication sessions in these networks. In particular, the problem of *minimum-energy broadcasting* in ad hoc wireless networks has gained a lot of attention recently [1,6,8,9,14,18,19,20]. It consists in designing a broadcast communication session which starts from a distinguished source node and minimizes the total energy consumption. The combinatorial difficulty of this problem follows from the observation that relaying signal between nodes

T. Erlebach (Ed.): CAAN 2006, LNCS 4235, pp. 112–124, 2006.

may result in lower transmission than communicating over large distance due to the nonlinear power attenuation, for an illustrating example see [19]. The problem of minimum energy broadcasting is known to be NP-hard both in its general graph version [10] and in its geometric version [6].

The geometric version of the problem where nodes are embedded in the two-dimensional Euclidean space has been quite extensively investigated. Three greedy heuristics have been proposed [20]: the minimum spanning tree (MST) heuristic, the shortest path tree (SPT) heuristic and the broadcasting incremental power (BIP) heuristic. These heuristics have been evaluated through simulations [20], and a quantitative characterization of their performances has been first performed by Wan et al. [19]. In particular, the approximation factor of the MST heuristic in the plane has been shown to be between 6 [19] and 12.15 [14]. In [6,7] Clementi et al. have shown among other things the first constant upper bounds and lower bounds (equal to the kissing number) on the approximation factor of the MST heuristic in the d-dimensional Euclidean space for $\alpha \geq d$. They have also observed that for $\alpha \leq d$, the approximation factor of the MST heuristic is $\Omega(n^{1-\alpha/d})$ and posed as an open problem the design of better approximation heuristics in this case. (Interestingly, for the related problem where the objective is to construct a minimum energy strongly connected network instead of a directed tree, the corresponding MST heuristic always yields at least 2-approximation [13].) Flammini et al. [9] have improved the aforementioned upper bounds for $\alpha \geq d$ to the form $3^d - 1$ and for $d = 2$ further down to 7.6. The factor of 7.6 was later further improved down to 6.33 by Navarra [18]. Recently, Ambühl [1] has shown that the MST heuristic in the plane approximates the optimum within the factor of 6 for $\alpha \geq 2$, improving the upper bound of 6.33 and matching the lower bound from [19]. Moreover, other heuristics than the aforementioned ones have been proposed. One of them is the *adaptive broadcast consumption* heuristic, which is related to MST but uses also backtracking in order to modify properly a current solution [14]. Another heuristic, relying on directed Steiner tree approach and achieving asymptotically an approximation factor of $O(n^\epsilon)$ with $\epsilon > 0$, has been provided by Liang [17].

Furthermore, even approximation algorithms with logarithmic ratios have been proposed for the more general graph version where the input is a complete directed graph $G = (V, E)$ with a symmetric cost function, i.e., $\text{cost}(u, v) = \text{cost}(v, u)$ for $u, v \in V$, associated with the edges in E. Caragiannis et al. [4] have provided an $10.8 \ln n$-approximation algorithm but this factor has been further improved down to $2 + 2\ln(n-1)$ by Calinescu et al. [3]. This has been proved to be asymptotically optimal [3].

We consider the d-dimensional geometric version of the minimum-energy broadcasting problem where the nodes are embedded in the d-dimensional Euclidean space in this paper. It is an important special case of the aforementioned general graph version occurring in practice [4]. In particular, finite three-dimensional wireless networks represent a wide category of practical networks, such as those deployed in a mountain terrain, air or water space, in buildings, or other three dimensional sensor networks [16].

We present the first non-trivial upper bounds on approximation factors of the three known greedy heuristics in the d-dimensional Euclidean space for $\alpha \leq d$. In particular, we show that the MST heuristic and the BIP heuristic approximate the optimum within $O(n^{1-\alpha/d})$ which asymptotically matches the lower bound for the MST heuristic observed in [6,7]. Next, we note that aforementioned lower bound is also valid for the BIP heuristic and show that for any $\alpha \geq 1$ (e.g., $\alpha \geq 3$) this heuristic in 3D can yield a solution at least $6+\frac{1}{3\alpha/2}+4\sin\frac{\pi}{12}$ times larger than the optimum. We also observe that the lower bound of $n/2$ on the approximation factor for the SPT heuristic from [19] generalizes to d-dimensions and arbitrary α and that the corresponding upper bound of $n-1$ holds. Furthermore, we consider the trivial heuristic where the whole broadcasting is done by the source node. We show that its approximation factor is lower bounded by $\Omega(n^{\alpha-1})$ and upper bounded by $n^{O(\alpha-1)}$ for any $\alpha \geq 1$.

The table in Fig. 1 summarizes the majority of our results, observations and known facts.

Heuristic	Lower Bound	Upper Bound
MST	$\Omega(n^{1-\alpha/d})$ [6,7]	$2^{O(d)}n^{1-\alpha/d}$
BIP	$\Omega(n^{1-\alpha/d})$	$2^{O(d)}n^{1-\alpha/d}$
SPT	$n/2$ [19]	$n-1$
trivial	$(n-1)^{\alpha-1}$	$n^{O(\alpha-1)}$

Fig. 1. Lower and upper bounds on the approximation factors of considered heuristics for minimum energy broadcasting in the d-dimensional Euclidean space for $1 \leq \alpha \leq d$

2   Preliminaries

Let d be a positive integer. We assume that the network nodes are given as a finite point set S in the d-dimensional Euclidean space. Let G be the complete graph over the point set S.

Any broadcast routing is a directed subtree T of G (termed *broadcasting tree*) rooted at the source node of the broadcasting that spans all nodes of G. We use $f_T(v)$ to denote the transmission power of the node v required by T. Then for any leaf node v of T, $f_T(v) = 0$; and for any internal node v of T,

$$f_T(v) = \gamma \cdot max_{vu\in T}|vu|^\alpha$$

where $|vu|$ stands for the Euclidean distance between u and v, i.e., the length of the segment uv, $\gamma > 0$, and $\alpha \geq 1$. The total energy required by T is then given by $\sum_{v\in T} f_T(v)$. The problem of minimum-energy broadcasting for G and a distinguished source node r of broadcasting in G is to find a directed tree T rooted at r which minimizes $\sum_{v\in T} f_T(v)$. Since the value of the constant γ does not affect the relative value of the broadcasting solution, we assume $\gamma = 1$ throughout the rest of the paper.

Following the definition by Wan et al. [19], we define the *radius* of a point set S as

$$\inf_{p \in S} \sup_{q \in S} |pq| .$$

For arbitrary weighted graphs, three greedy heuristics for the minimum-energy broadcasting are well known [19,20]. The MST heuristic constructs a MST of the graph and then it roots the tree at the source node and directs its edges. The SPT heuristic constructs first, for each node v, a minimum-cost path from the source node to v by applying, e.g., Dijkstra's algorithm. Note that a cost, involving power, can be defined for each link in the network. The broadcast tree consists of the superposition of these unicast paths. The BIP heuristic can be seen as a variant of Dijkstra's algorithm. It grows a directed tree rooted at the source node, starting from the source node. New nodes are added on a minimum incremental cost basis until all nodes are spanned by the tree.

3   Lower Bounds

Clementi et al. generalized the lower bound of 6 for the MST heuristic in the plane from [19] to include higher dimensions, as follows in [7].

Fact 1. *For any $\alpha \geq 1$, the approximation factor of the MST heuristic in the d-dimensional Euclidean space is not less than the kissing number in this space (e.g., 12 for $d = 3$).*

In the following, we show that the lower bound of $\frac{13}{3}$ for the BIP heuristic in the two-dimensional Euclidean space from [19] can be also generalized to higher dimensions, specifically to 3D.

Theorem 1. *For any $\alpha \geq 1$, the approximation factor of the BIP heuristic in the three-dimensional Euclidean space is at least $6 + \frac{1}{3^{\alpha/2}} + 4\sin\frac{\pi}{12}$. In particular, for $\alpha = 1$, the approximation factor is at least $6 + \frac{2}{\sqrt{3}} + 4\sin\frac{\pi}{12}$.*

Proof. Sketch. Consider a ball B of radius 1 centered at the source node r. Let H be a hyper-plane including r and let C be the unit circle which is the intersection of H with the surface of B. For any three points p, q and t, the angle between the two rays pq and pt is denoted by $\angle qpt$ or $\angle tpq$.

First, let us construct the point nodes $q_1, ..., q_{m+1}, p_1, ..., p_6$ within C on H as in the proof of the $\frac{13}{3}$ lower bound for BIP in the planar case in [19] (see Fig. 2).

Namely, let θ be a sufficiently small positive number. The angles between the point nodes satisfy the following equalities:

$$\angle p_1 r p_2 = \angle p_5 r p_6 = \frac{\pi}{3} - 3\theta ,$$

$$\angle p_2 r p_3 = \angle p_4 r p_5 = \frac{\pi}{3} - 2\theta ,$$

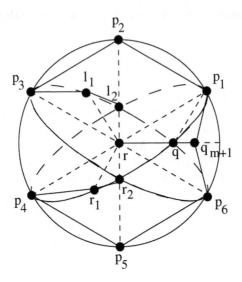

Fig. 2. A bad instance for BIP in 3D

$$\angle p_3 r p_4 = \frac{\pi}{3} - \theta \ ,$$

$$\angle p_6 r p_1 = \frac{\pi}{3} + 11\theta \ .$$

Next, let q be the point in the perpendicular bisector of $p_1 p_6$ so that $p_1 q$ is perpendicular to $p_1 p_2$. Let m be sufficiently large to satisfy the inequality $1 - (\frac{|rq|}{m})^2 > |p_3 p_4|^2$. Then the nodes $q_1, ..., q_{m+1}$ are the $m+1$ points on the ray rq satisfying $|rq_i| = \frac{i}{m}|rq|$ for $1 \leq i \leq m+1$. Note that $q_m = q$.

Further, we extend the construction of [19] to a three-dimensional one as follows.

For $i \in \{3, 4\}$, let H_i be the hyper-plane perpendicular to H including r and p_i. The hyper-plane H splits B into, say, a left and right half-ball. Let C_3 be the half-circle on the intersection of H_3 with the surface of the left half-ball and similarly let C_4 be the half-circle on the intersection of H_4 with the right half-ball. Place points l_1 and l_2 on C_3 so that angles $\angle p_3 r l_1$ and $\angle l_1 r l_2$ are respectively $\frac{\pi}{3} - \frac{2}{3}\theta$ and $\frac{\pi}{6} - \frac{\theta}{3}$. Symmetrically, place points r_1 and r_2 on C_4 so that angles $\angle p_4 r r_1$ and $\angle r_1 r r_2$ are respectively $\frac{\pi}{3} - \frac{2}{3}\theta$ and $\frac{\pi}{6} - \frac{\theta}{3}$.

By the construction, for sufficiently large m and sufficiently small θ, in the first $m+1$ steps the BIP heuristic adds the points $q_1, \ldots, q_{m+1}$, and the transmission power of the nodes $r, q_1, \ldots, q_m$ is $\frac{m+1}{m^\alpha}|rq|^\alpha$. In the next step, the points p_1 and p_6 are added. Note that they are closer to any q_i point, in particular, q_{m+1}, than l_2 or r_2. Then, analogously as in [19], the points p_2, p_5, p_3 and p_4 are added one by one in four steps.

By the straightforward generalization of the proof in [19], the first $m + 6$ steps use

$$\frac{m+1}{m^\alpha}|rq|^\alpha + |p_1q_{m+1}|^\alpha + 2|p_1p_2|^\alpha + 2|p_2p_3|^\alpha$$

energy.

In the next subsequent two steps, the points l_1 and r_1 are added. Finally, in the last subsequent two steps, the points l_2 and r_2 are added. Totally, the last four steps require $|p_3l_1|^\alpha + |p_4r_1|^\alpha + |l_1l_2|^\alpha + |r_1r_2|^\alpha$ energy.

Since as observed in [19], for θ tending to 0 and m tending to infinity, $|p_1q_{m+1}|$ tends to $\frac{1}{\sqrt{3}}$, the total energy used by BIP for $\alpha > 1$ tends to $\frac{1}{3^{\alpha/2}} + 4 + 2 +$ $4\sin\frac{\pi}{12}$, i.e., $6 + \frac{1}{3^{\alpha/2}} + 4\sin\frac{\pi}{12}$. Note that for $\alpha = 1$ the total energy is at least $6 + \frac{2}{\sqrt{3}} + 4\sin\frac{\pi}{12}$.

On the other hand, a broadcasting from the source r requiring 1^3 energy is sufficient and hence, the theorem follows.   □

The approximation behavior of the MST heuristic is known to change dramatically when α becomes smaller than d. Clementi et al. observed in [6,7] that for $\alpha \leq d$ the MST heuristic applied to the set of n grid points in the d-dimensional integer grid $n^{1/d} \times .... \times n^{1/d}$ yields a solution requiring $\Omega(n)$ energy whereas the outcome of the trivial heuristic requires only $O(n^{\alpha/d})$ energy. Hence, we have the following fact.

Fact 2. *For any $\alpha \leq d$, the approximation factor of the MST heuristic in the d-dimensional Euclidean space is $\Omega(n^{1-\alpha/d})$.*

The observation of Clementi et al. is easily seen to be valid also for the BIP heuristic.

Remark 1. *For any $\alpha \leq d$, the approximation factor of the BIP heuristic in the d-dimensional Euclidean space is $\Omega(n^{1-\alpha/d})$.*

The proof of the lower bound $n/2$ on the approximation factor of the SPT heuristic in two dimensions given in [19] can be immediately adapted to $d \geq 3$ dimensions. Hence, we obtain

Remark 2. *For any $\alpha \geq 1$, the approximation factor of the SPT heuristic in the d-dimensional Euclidean space is at least $n/2$.*

Finally, let us examine the trivial heuristic where solely the source node broadcasts.

Remark 3. *For any $\alpha \geq 1$, the approximation factor of the trivial heuristic in the d-dimensional Euclidean space is at least $(n-1)^{\alpha-1}$.*

Proof. Place the n input points on a unit length segment so the source point lies at the endpoint of the segment and the distance between two consecutive points is $\frac{1}{n-1}$. The trivial heuristic uses 1^α, i.e., 1 energy whereas the optimal solution is easily seen to require only $(n-1) \times \frac{1}{(n-1)^\alpha}$, i.e., $(n-1)^{1-\alpha}$ energy.   □

4 Upper Bounds

In this section, we provide tight upper bounds on the approximation factors of the four considered heuristics in the d-dimensional Euclidean space (for $\alpha \leq d$), asymptotically matching the corresponding lower bounds presented in the preceding section.

Let T be a MST of the input point set S in the d-dimensional Euclidean space. Next, let c denote the supreme of $\sum_{e \in T} |e|^\alpha$ over all point sets S of radius one. The proof of the following lemma is completely analogous to the proof of the corresponding lemma for a planar point set given in [19].

Lemma 1. *For any $\alpha \geq 1$ and any point set S in the d-dimensional Euclidean space, the total energy required by any broadcasting among S is at least $\frac{1}{c} \sum_{e \in T} |e|^\alpha$.*

In order to make the analysis more self-contained, we present a complete and independent analysis of the MST heuristic although the result of an initial fragment of our analysis, i.e., the inequality $\sum_{e \in T} |e|^d \leq 2^{O(d)}$ for a MST T in the d-dimensional Euclidean space is known in the literature on MST for $\alpha \geq d$, e.g., see [9,18]. Our proof of the aforementioned fragment (mostly given in Lemma 2) is quite simple.

Associate with each edge e of T the d-dimensional ball $B(e)$ of radius $\epsilon |e|$ centered at the midpoint of e.

Lemma 2. *For $\epsilon < (1 - 1/\sqrt{2})/2$, the balls associated with the edges of T are pairwise disjoint.*

Proof. Suppose that $B(e_1) \cap B(e_2) \neq \emptyset$ for two distinct edges e_1 and e_2 in T. It follows that the segment s connecting the midpoints of e_1 and e_2 is of length not greater than $2\epsilon \cdot \max\{|e_1|, |e_2|\}$. Consider the hyper-plane H induced by e_2 and the segment s. Let e'_2 be a segment parallel to e_2 on H and of the same length as e_2 such that the midpoint of e'_2 is the other endpoint of s. Note that e_1 and e'_2 cross each other in their midpoints and the endpoints of e'_2 are within $2\epsilon \cdot \max\{|e_1|, |e_2|\}$ distance from the corresponding endpoints of e_2.

Consider the hyper-plane H' induced by e_1 and e'_2. The endpoints of e_1 and e'_2 induce a parallelogram P whose diagonals are e_1 and e'_2 in H. We may assume w.l.o.g the configuration of Fig. 3 where the base b of the parallelogram is not shorter than its side a and e_1 is not shorter than e'_2.

Suppose first that there is a path in T which connects the left endpoint of e_1 with the endpoint of e_2 corresponding to the right endpoint of e'_2 and does not use e_1 or e_2. Then, if we remove e_1 from T and insert an edge e_3 connecting the right endpoint of e_1 with the endpoint of e_2 corresponding to the right endpoint of e'_2, we obtain another spanning tree of S whose length is not greater than $|T| - |e_1| + |e_3|$.

By the relationship $2(|a|^2 + |b|^2) = |e_1|^2 + |e'_2|^2$ between the diagonals and the sides of the parallelogram P and our assumptions on them, we obtain $4|a|^2 \leq 2|e_1|^2$ and consequently $|a| \leq \frac{|e_1|}{\sqrt{2}}$. Hence, the inequality $|e_3| \leq |e_1|/\sqrt{2} + 2\epsilon |e_1|$

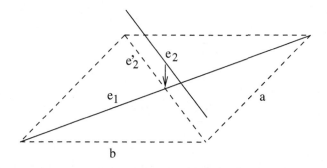

Fig. 3. The parallelogram P

holds. This implies $|e_3| < |e_1|$ for $\epsilon < (1 - 1/\sqrt{2})/2$ which is a contradiction with the optimality of T.

Suppose in turn that there is a path in T which connects the right endpoint of e_1 with the endpoint of e_2 corresponding to the right endpoint of e'_2 and does not use e_1 or e_2. Then, if we remove e_1 from T and insert an edge e_4 connecting the left endpoint of e_1 with the endpoint of e_2 corresponding to the left endpoint of e'_2, we obtain another spanning tree of S whose length is not greater than $|T| - |e_1| + |e_4|$. We obtain similarly a contradiction for $\epsilon < (1 - 1/\sqrt{2})/2$ by $|e_4| \leq |e_1|/\sqrt{2} + 2\epsilon|e_1|$ and the optimality of T.

The two remaining cases of the connecting path are symmetric to the considered ones. □

Lemma 3. *For any $\alpha \in [1, d]$ and any point set S of radius one in the d-dimensional Euclidean space and its MST T, the equality*

$$\sum_{e \in T} |e|^{\alpha} = O(n^{1 - \frac{\alpha}{d}})$$

holds.

Proof. Observe that since S is of radius one, all the edges of T have length not greater than 1. It follows also that for $\epsilon < (1 - 1/\sqrt{2})/2$, the total volume of the balls $B(e)$ over the edges e in T is

$$\epsilon^d f(d) \sum_{e \in T} |e|^d$$

where f is a function of d, e.g., $f(3) = \frac{4\pi}{3}$, see [11] for the general form of f. On the other hand, all the balls $B(e)$ are within a ball of radius $1 + \epsilon$. Hence, we obtain

$$\epsilon^d f(d) \sum_{e \in T} |e|^d \leq (1 + \epsilon)^d f(d) .$$

Therefore,

$$\epsilon^d \sum_{e \in T} |e|^d \leq (1 + \epsilon)^d ,$$

yielding

$$\sum_{e \in T} |e|^d \leq 2^{O(d)}$$

when ϵ tends to $(1 - 1/\sqrt{2})/2$.

We may assume without loss of generality that n is a power of two. For $i = 1, ..., \log n$, let m_i be the number of edges in T whose length is in the interval $[\frac{2^{i-1}}{n}, \frac{2^i}{n})$. It follows that $m_i \frac{2^{d(i-1)}}{n^d} \leq 2^{O(d)}$ and consequently $m_i \leq 2^{O(d)} 2^{-d(i-1)} n^d$. Next, let m_0 be the number of edges in T of length less than $\frac{1}{n}$. Clearly, we have $\sum_{i=0}^{\log n} m_i \leq n$. Hence, on the other hand, we obtain

$$\sum_{e \in T} |e|^\alpha \leq O(n \times (\frac{2^{(d-1)\log n/d}}{n})^\alpha) + \sum_{i=\lceil (d-1)\log n/d \rceil}^{\log n} m_i \frac{2^{\alpha i}}{n^\alpha} .$$

Combining these inequalities, we obtain

$$\sum_{e \in T} |e|^\alpha \leq O(n^{1-\alpha/d}) + 2^{O(d)} n^{d-\alpha} \sum_{i=\lceil (d-1)\log n/d \rceil}^{\log n} 2^{\alpha i - d(i-1)} .$$

Consequently, we have

$$\sum_{e \in T} |e|^\alpha = O(n^{1-\alpha/d}) + 2^{O(d)} \sum_{i=\lceil (d-1)\log n/d \rceil}^{\log n} 2^{\alpha(i-\log n)+d(\log n - i+1)}$$

$$= O(n^{1-\alpha/d}) + 2^{O(d)} O(\sum_{i=\lceil (d-1)\log n/d \rceil}^{\log n} 2^{(d-\alpha)(\log n - i)}) .$$

Thus, we obtain

$$\sum_{e \in T} |e|^\alpha = O(n^{1-\alpha/d}) + 2^{O(d)} O(\sum_{j=1}^{\lfloor \log n/d \rfloor} 2^{(d-\alpha)j})$$

$$= O(n^{1-\alpha/d}) + 2^{O(d)} O(\sum_{j=1}^{\lfloor \log n/d \rfloor} 2^j)^{(d-\alpha)}$$

$$\leq 2^{O(d)} n^{1-\alpha/d} .$$

$\square$

By combining Lemma 1 with Lemma 3, we obtain our first main result.

Theorem 2. *For any* $\alpha \in [1, d]$, *the MST heuristic in the d-dimensional Euclidean space yields an* $2^{O(d)}n^{1-\alpha/d}$-*approximation.*

Wan et al. observed in [19] that the total energy required by the broadcasting tree constructed by the BIP heuristic for a planar point set is at most $\sum_{e \in T} |e|^\alpha$ where T is a MST of the input point set. Their proof works for a point set in the d-dimensional Euclidean space as well. Hence, we have the following lemma.

Lemma 4. *The total energy required by the broadcasting tree constructed by the BIP heuristic for a point set in the d-dimensional Euclidean space is at most* $\sum_{e \in T} |e|^\alpha$ *where T is a MST of the input point set.*

Combining Lemma 4 with Lemma 3, we obtain our bound for the BIP heuristic.

Theorem 3. *For any* $\alpha \in [1, d]$, *the BIP heuristic in the d-dimensional Euclidean space yields an* $O(n^{1-\alpha/d})$-*approximation.*

Note that the energy required by the SPT heuristic does not exceed $n - 1$ times the maximum of the minimum energy required by routing from the source node r to a node v taken over all the nodes v different from r. Hence, we immediately obtain the following remark.

Remark 4. *For any* α, *the approximation factor of the SPT heuristic in any d-dimensional Euclidean space is not greater than* $n - 1$.

In turn, let us consider the trivial heuristic where the source node performs the whole broadcast. It uses r^α energy where r is the radius of the input point set. To analyze it we shall use the following algebraic lemma.

Lemma 5. *For any* $\alpha \geq 1$ *and positive* a, b,

$$(a + b)^\alpha \leq 2^{\alpha-1}(a^\alpha + b^\alpha) .$$

Proof. The inequality is equivalent to the following one

$$((a + b)/2)^\alpha \leq (a^\alpha + b^\alpha)/2 .$$

Consider the power function x^α. The left side of the inequality is equal to the value of the function at the midpoint between a and b. In turn, the right side of the inequality is equal to the mean of the values of the function at a and b, respectively. It remains to observe that the power function is convex for $\alpha \geq 1$. □

Theorem 4. *For any* α, *the approximation factor of the trivial heuristic in any d-dimensional Euclidean space is not greater than* $n^{O(\alpha-1)}$.

Proof. Consider the directed tree rooted at the source node inducing a minimum energy broadcast session. Suppose first that the tree is just a simple directed path. Let (u, v), (v, w) be two incident edges on the path. If we replace them by the edge (u, w), the corresponding broadcast session will use $dist(u, w)^\alpha -$

$dist(u, v)^\alpha - dist(v, w)^\alpha$ more energy. By the two inequalities $dist(u, w)^\alpha \leq (dist(u, v) + dist(v, w))^\alpha$ and $(dist(u, v) + dist(v, w))^\alpha \leq 2^{\alpha-1}(dist(u, v)^\alpha + dist(v, w)^\alpha)$, the increase in energy does not exceed $(2^{\alpha-1} - 1) \cdot (dist(u, v)^\alpha + dist(v, w)^\alpha)$. We can easily divide the path into pairs of consecutive edges and at most one single edge. It follows that replacing each such a pair with a single edge increases energy consumption by a multiplicative factor not greater than $2^{\alpha-1}$. After $O(\log n)$ edge replacing phases the path reduces to a single edge, say e, leaving r and the tree transforms into a star centered at r where e is the longest edge. By induction on the number of phases and the observation that $(2^{\alpha-1})^{O(\log n)} = n^{O(\alpha-1)}$, the broadcast session induced by the resulting star uses at most $n^{O(\alpha-1)}$ times more energy than that used by the optimal tree.

Consider the general case where the optimal tree is not necessarily a path. Let U be the total energy used by the optimal tree. Observe that for any path from r to a leaf in the tree, the sum of the lengths of its edges raised to the α power does not exceed U. Hence, if we assign $n^{O(\alpha-1))}U$ energy to r (with the same constant at $\alpha - 1$ in the exponent as in the path case) any vertex on any such a path will be within the range of r by the argumentation from the path case. $\qquad\square$

5   Concluding Remarks

We have provided a tight upper bound on the approximation factor of the MST heuristic for minimum energy broadcasting in the d-dimensional Euclidean space for $\alpha < d$ matching the lower bound observed by Clementi et al. in [6,7] as well as tight upper and lower bounds for three other standard heuristics for minimum energy broadcasting adopted to d-dimensional Euclidean space and $\alpha < d$.

It is an intriguing question whether or not the problem of minimum energy broadcasting in the d-dimensional Euclidean space for $\alpha < d$ admits an $o(\log n)$ approximation in polynomial time.

Acknowledgments

The authors are very grateful to Anna Torstensson for shortening the proof of Lemma 5, Andrea Clementi for useful information and Joachim Gudmundsson for valuable discussions.

References

1. C. Ambühl. An optimal bound for the MST algorithm to compute energy efficient broadcast trees in wireless networks. In *Proc. 32nd International Colloquium on Automata, Languages and Programming (ICALP 2005)*, volume 3580 of LNCS, pages 1139–1150, Springer-Verlag, 2005.
2. H. Bronnimann and M. T. Goodrich. Almost Optimal Set Covers in Finite VC-Dimension. In *Proc. 10th Annual ACM Symposium on Computational Geometry (SCG'94)*, pages 293–302, ACM Press, 1994.

3. G. Calinescu, S. Kapoor, A. Olshevsky and A. Zelikovsky. Network Lifetime and Power Assignment in ad hoc Wireless Networks. In *Proc. 11th Annual European Symposium on Algorithms (ESA'03)*, volume 2832 of LNCS, pages 114–126, Springer-Verlag, 2003.
4. I. Caragiannis, C. Kaklamanis and P. Kanellopoulos. New results for energy-efficient broadcasting in wireless networks. In *Proc. 13th Annual International Symposium on Algorithms and Computation (ISAAC'02)*, volume 2518 of LNCS, pages 332–343, Springer-Verlag, 2002.
5. K. L. Clarksson and K. Varadarajan. Improved Approximation Algorithms for Geometric Set Cover. In *Proc. 21st Annual ACM Symposium on Computational Geometry (SCG'05)*, pages 135–141, ACM Press, 2005.
6. A. E. F. Clementi, P. Crescenzi, P. Penna, G. Rossi, and P. Vocca. On the complexity of computing minimum energy consumption broadcast subgraphs. In *Proc. 18th Annual Symposium on Theoretical Aspects of Computer Science (STACS'01)*, volume 2010 of LNCS, pages 121–131, Springer-Verlag, 2001.
7. A. E. F. Clementi, P. Crescenzi, P. Penna, G. Rossi, and P. Vocca. A Worst-case Analysis of an MST-based Heuristic to Construct Energy-Efficient Broadcast Trees in Wireless Networks. Manuscript.
8. A. E. F. Clementi, G. Huiban, P. Penna, G. Rossi, and Y. C. Verhoeven. Some Recent Theoretical Advances and Open Questions on Energy Consumption in Ad-Hoc Wireless Networks. In *Proc. 3rd Workshop on Approximation and Randomization Algorithms in Communication Networks (ARACNE 2002)*, pages 23–38, 2002.
9. M. Flammini, R. Klasing, A. Navarra, and S. Prennes. Improved approximation results for the minimum energy broadcasting problem. In *Proc. of the 2004 joint workshop on Foundations of mobile computing (DIALM-POMC 04)*, pages 85–91, ACM Press, 2004.
10. M. R. Garey and D. S. Johnson. Computers and Intractability: a Guide to the Theory of NP-Completeness. W.H. Freeman and Company, 1979.
11. J. E. Goodman, and J. O'Rourke (eds.). Handbook of Discrete and Computational Geometry. Chapman & Hall/CRC, 2004.
12. Z. Haas and S. Tabrizi. On Some Challenges and Design Choices in Ad-Hoc Communications. In *Proc. of the IEEE Military Communication Conference (MILCOM 98)*, pages 187–192, 1998.
13. L. M. Kirousis, E. Kranakis, D. Krizanc, and A. Pelc. Power Consumption in Packet Radio Networks. Theoretical Computer Science 243, pages 289–305, 2000.
14. R. Klasing, A. Navarra, A. Papadopoulos, and S. Prennes. Adaptive broadcast consumption (ABC), a new heuristic and new bounds for the minimum energy broadcast routing problem. In *Proc. 3rd International IFIP-TC6 Networking Conference (Networking 2004)*, volume 3042 of LNCS, pages 866–877, Springer-Verlag, 2004.
15. G. S. Lauer. *Packet radio routing*, chapter 11 of *Routing in communication networks*, M. Streenstrup (ed.), pages 351-396. Prentice-Hall, 1995.
16. G. Li, P. Fan, and K. Cai. On the geometrical characteristics of three dimensional wireless ad-hoc networks and its applications. To appear in EURASIP Journal on Wireless Communications & Networking.
17. W. Liang. Constructing minimum-energy broadcast trees in wireless ad hoc networks. In *Proc. 3rd ACM International Symposium on Mobile Ad Hoc Networking & Computing (MobiHoc 2002)*, pages 112–122, ACM Press, 2002.
18. A. Navarra. Tighter bounds for the minimum energy broadcasting problem. In *Proc. 3rd International Symposium on Modeling and Optimization in Mobile, Ad Hoc, and Wireless Networks (WiOpt 2005)*, pages 313–322, IEEE Computer Society, 2005.

19. P. J. Wan, G. Calinescu, X. Y. Li, and O. Frieder. Minimum-energy broadcast routing in static ad hoc wireless networks. In *Proc. 20th Annual Joint Conference of the IEEE Computer and Communications Societies (INFOCOM 2001)*, pages 1162–1171, 2001.
20. J. E. Wieselthier, G. D. Nguyen, and A. Ephremides. On the construction of energy-efficient broadcast and multicast trees in wireless networks. In *Proc. 19th Annual Joint Conference of the IEEE Computer and Communications Societies (INFO-COM 2000)*, pages 585–594, 2000.

Optimal Gossiping with Unit Size Messages in Known Topology Radio Networks*

Fredrik Manne and Qin Xin

Department of Informatics, The University of Bergen, Norway
{fredrikm, xin}@ii.uib.no.

Abstract. Gossiping is a communication primitive where each node of a network possesses a unique message that is to be communicated to all other nodes in the network. We study the gossiping problem in known topology radio networks where the schedule of transmissions is precomputed in advance based on full knowledge about the size and the topology of the network. In addition we consider the case where it is only possible to transmit a unit size message in each time step. This gives a more realistic model than if arbitrary length messages can be sent during each time step, as has been the case in most previous studies of the gossiping problem. In this paper, we propose an optimal randomized schedule that uses $O(n \log n)$ time units to complete the gossiping task with high probability in any radio network of size n. This matches the lower bound of $\Omega(n \log n)$ by Gąsieniec and Potapov in [17] [TCS'02]. Our new gossiping schedule is based on the notion of a gathering spanning tree proposed by Gąsieniec, Peleg and Xin in [19] [PODC'05].

Keywords: Centralized radio networks, gossiping, randomized schedule.

1 Introduction

The two classical problems of information dissemination in computer networks are the *broadcasting* problem and the *gossiping* problem. The broadcasting problem requires distributing a particular message from a distinguished *source* node to all other nodes in the network. In the gossiping problem, each node v in the network initially holds a message m_v, and the aim is to distribute all messages to all nodes. For both problems, one generally considers as the efficiency criterion the minimization of the time needed to complete the task.

This paper concerns the following model of a radio network. A network is an undirected connected graph $G = (V, E)$, where V represents the set of nodes of the network and E contains unordered pairs of distinct nodes, such that $(v, w) \in E$ iff the transmissions of node v can directly reach node w and vice versa (the reachability of transmissions is assumed to be a symmetric relation). In this case, we say that the nodes v and w are *neighbours* in G. One of the particular properties of radio network is that a message transmitted by a node is always sent to all of its neighbours.

---

* Supported by the Research Council of Norway through the SPECTRUM project.

T. Erlebach (Ed.): CAAN 2006, LNCS 4235, pp. 125–134, 2006.

The number of neighbours of a node w is called its *degree*, and the maximum degree of any node in the network is called the *maximum degree* of the network and is denoted by Δ. The *size of the network* is the number of nodes $n = |V|$.

Communication in the network is synchronous and consists of a sequence of communication steps. During each step, each node v either transmits or listens. If v transmits, then the transmitted message reaches each of its neighbours by the end of this step. However, a node w adjacent to v successfully receives this message iff w is listening during this step and v is the only transmitting node among w's neighbours. If node w is adjacent to a transmitting node but it is not listening, or it is adjacent to more than one transmitting node, then a *collision* occurs and w does not retrieve any message in this step.

The running time of any communication schedule is determined by the number of time steps required to complete the communication task. That is, we do not account for any internal computation within individual nodes.

Most of the work in this field has been done under the assumption that the processors can transmit messages of arbitrary size in a single time step. In particular any node can send all of the information it has received so far in a single (atomic) step of the communication process. Note that this strong assumption is rather unrealistic if the size of the network is very large. In this paper we study the gossiping problem in radio networks where there is a restriction on the size of each message. In particular we investigate the case where each message is of unit size, meaning that it contains information originating from exactly one node of the network.

We focus on algorithms that rely on using complete information about the network topology. This type of topology-wise communication algorithms are useful in radio networks that have a reasonably stable topology/infrastructure. As long as no changes occur in the network topology during the execution of the algorithm, the tasks of broadcasting and gossiping will be completed successfully. Note also that our main goal is the design of time efficient communication procedures. However, it would not be difficult to increase the level of fault-tolerance in our algorithm at the expense of some small extra time consumption.

Communication in radio networks with known topology. The work on communication in known topology radio networks was initiated in the context of the broadcasting problem. In [13], Gaber and Mansour prove that the broadcasting task can be completed in time $O(D \log^2 n)$ where D is the diameter of the network. An $\Omega(\log^2 n)$ time lower bound was proved for the family of graphs of radius 2, see [2] by Alon *et al.* While it was known for quite a while that for every n-node radio network that there exists a deterministic broadcasting schedule of length $O(D \log n + \log^2 n)$, Bar-Yehuda *et al.* [3], an appropriate efficient construction for such a schedule was only recently proposed in [21] by Kowalski and Pelc. Subsequently, an efficient deterministic construction of a broadcasting schedule of length $D + O(\log^4 n)$ was proposed by Elkin and Kortsarz [12]. In this paper, they also present an efficient deterministic construction for a broadcasting schedule of length $D + O(\log^3 n)$ for planar graphs. In [19], Gąsieniec, Peleg and Xin proposed a more efficient deterministic schedule that uses $D + O(\log^3 n)$ time units to complete the broadcasting task in any radio network. This paper also contains an optimal randomized broadcasting schedule of length $D + O(\log^2 n)$ and a new broadcasting schedule using fewer than $3D$ time slots on planar graphs. More recently in [7], Cicalese, Manne and Xin improved the broadcasting time to $D + O(\frac{\log^3 n}{\log \log n})$ in any radio network.

Efficient radio broadcasting algorithms for several types of known network topologies can be found in Diks *et al.* [10]. For general networks, however, it is known that the computation of an optimal (radio) broadcast schedule is NP-hard, even if the underlying graph is embedded in the plane [5,23].

Radio gossiping in networks with known topology was first studied in the context of radio communication with messages of limited size, see [17] by Gąsieniec and Potapov. In this model the authors proposed several optimal or close to optimal $O(n)$-time gossiping procedures for various standard network topologies, including lines, rings, stars and free trees. They also proved that there exists a radio network topology in which the gossiping (with unit size messages) requires $\Omega(n \log n)$ time. The first work on radio gossiping in known topology networks with arbitrarily large messages is [18], where Gąsieniec, Potapov and Xin propose several optimal gossiping schedules for a wide range of radio network topologies. Very recently, Gąsieniec, Peleg and Xin proposed an efficiently computable deterministic schedule that uses $O(D + \Delta \log n)$ time units to complete the gossiping task in any radio network [19]. This improves on the previous best known gossiping schedule [18] with running time $O(D + {}^{i+2}\!\sqrt{D}\Delta \log^{i+1} n)$, for any network with diameter $D = \Omega(\log^{i+4} n)$, where i is an arbitrary integer constant $i \geq 0$. Subsequently in [7], Cicalese, Manne and Xin improved the gossiping time even further to $O(D + \frac{\Delta \log n}{\log \Delta - \log \log n})$ in radio networks where $\Delta = \Omega(\log n)$.

Our results. In this paper, we study the gossiping problem in known topology radio networks, where during each time step only one unit size message originating from some node of the network can be transmitted successfully. The schedule of transmissions is precomputed in advance based on full knowledge about the size and the topology of the network. We propose an optimal randomized schedule that uses $O(n \log n)$ time units to complete the gossiping task with high probability in any radio network of size n. This matches the lower bound of $\Omega(n \log n)$ by Gąsieniec and Potapov in [17]. Our new gossiping schedule is based on the notion of a gathering spanning tree proposed by Gąsieniec, Peleg and Xin in [19].

2 Centralized Gossiping with Unit Size Messages in Arbitrary Graphs

In this section, we study the time complexity of gossiping in general undirected graphs. We show that radio gossiping with unit size messages in undirected graphs can be performed in time $O(n \log n)$ with high probability. Our gossiping algorithm runs in two stages. In the first stage, we collect all the messages in a distinguished central node c by transporting messages along branches of any BFS spanning tree rooted in c. The second stage is performed through broadcasting of n unit messages from c. These broadcasts are performed in a pipelined fashion along a gathering spanning tree, a structure first proposed by Gąsieniec, Peleg and Xin in [19].

2.1   Gathering Messages in Arbitrary Graphs

Given an arbitrary graph $G = (V, E)$ and a BFS spanning tree T rooted at its central node c, we partition the nodes into consecutive layers $L_i = \{v \mid dist(c, v) = i\}$, for $i = 0, .., r$ where r is a radius of T.

In the following, we will use the standard notions of *parent, children, and descendant* in trees. For simplicity we assume that a node is a descendant of itself.

We say that a node v is unsecured iff v has not delivered all messages stored originally in its descendant to $parent(v)$ in T. The different messages are transmitted toward c in a pipelined fashion, such that only one unsecured node v in L_i of T is allowed to transmit a message to $parent(v)$ in T at any time step t, for $1 \leq i \leq r$. To avoid collisions between different BFS layers, the nodes in L_i are allowed to transmit in time steps t satisfying $t = i \bmod 3$.

The following result follows directly.

Lemma 1. *All different messages can be gathered at the central nodes c of G in time $3(n - 1)$.*

2.2   Broadcasting in Arbitrary Graphs

Now that all messages have been gathered in c, we will show how we with high probability can broadcast them to all nodes in G in $O(n \log n)$ time.

We first recall the following recursive ranking procedure of nodes in a tree (see [19]). Leaves have rank 1. Next consider a node v and the set Q of its children and let r_{max} be the maximum rank of the nodes in Q. If there is a unique node in Q of rank r_{max} then set the rank of v to r_{max}, otherwise set the rank of v to $r_{max} + 1$.

Lemma 2. *The largest rank in a tree of size n is bounded by $\lceil \log n \rceil$. (see [19]).*

Given any graph G with central node c, a *gathering spanning* tree of G is a BFS spanning tree T of G rooted at c, such that T is ranked as above and that also satisfies the following condition: every node in L_{j+1} of rank i is at most adjacent to one node in L_j also of rank i, and if all the nodes of rank i in L_j transmit at the same time then the messages will be received by the nodes in L_{j+1} of rank i successfully without any collision.

The following lemma was shown in [19].

Lemma 3. *There exists a polynomial time construction of a gathering spanning tree in any graph G.*

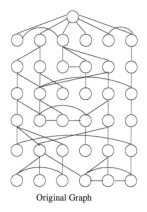
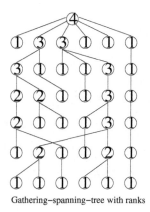

Original Graph   Gathering–spanning–tree with ranks

Fig. 1. Creating a gathering spanning tree

Figure 1 shows how a gathering spanning tree can be constructed from a graph G.

For a gathering spanning tree T we say that an edge in T is *fast* if both of its end points have the same rank, and it is *slow* otherwise. Since the largest rank is at most $\lceil \log n \rceil$, there are at most $\lceil \log n \rceil$ slow edges in each path from the root c to any leaf of T.

For a graph G with gathering spanning tree T, we now partition the edges of every path of T from c to a leaf into consecutive edge segments in the following way.

(1) Every maximal connected path of fast edges in T is a *fast* segment.
(2) Every slow edge is a *slow* segment.

Note that a node can belong to both a fast and a slow segment. In the following, the time steps t are divided into *fast blocks* ($t = 0 \bmod 2$) and *slow blocks* ($t = 1 \bmod 2$), such that the communication within the fast segments of T only occur in the fast blocks and similarly, communication within the slow segments of T only occur in the slow blocks. We will not be explicit about this schedule in the future presentation but assume that the time units used for both the fast and slow segments are consecutive.

We now define a graph $\tilde{G} = (\tilde{V}, \tilde{E})$ as follows. Its nodes are the same as in G and $E \subset \tilde{E}$. In addition for every node v in a fast segment of T we add an edge (v, w) where w is the topmost node of the fast segment that v belongs to. See Figure 2 for an example.

Lemma 4. *The graph $\tilde{G}$ has radius at most $2 \log n$.*

Proof. The lemma follows directly from the definition of $\tilde{G}$ and Lemma 2.2.

To simplify our presentation, we also define a modified gathering spanning tree $\tilde{T}$ of $\tilde{G}$ as follows. The central node c is the root of $\tilde{T}$ as well. Every edge of T between two nodes of different rank will be also an edge in $\tilde{T}$. In addition, if w

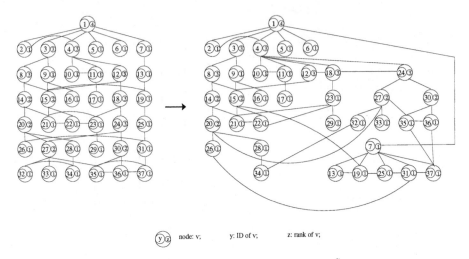

Fig. 2. From the original graph G to $\tilde{G}$

is the topmost node of a fast segment F in T, then w has an adjacent edge in $\tilde{T}$ to each node $v \in F$. We denote each node that belongs to a fast segment in T as fast in $\tilde{T}$. Thus the main change from T to $\tilde{T}$ is that the nodes of every fast segment in T has been collapsed and are now hanging of the topmost node in the segment. For all other nodes their parent relationship in $\tilde{T}$ is the same as in T. The main purpose for defining $\tilde{T}$ is to be able to reason about the time complexity of performing the slow transmissions.

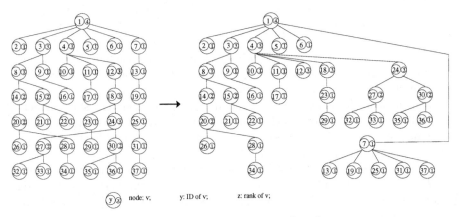

Fig. 3. From a T of G to a $\tilde{T}$ of $\tilde{G}$

Observation 1. $\tilde{T}$ of $\tilde{G}$ spans G as well.

We are now ready to describe the gossiping schedule, which consists of *fast steps* and *slow steps* as described below.

(1) **Fast steps:** These are used for transmissions between two fast nodes that are adjacent in T. Let $r_{max} \leq \lceil \log n \rceil$ be the largest rank in T. Consider a node v of rank j, $1 \leq j \leq r_{max}$ on BFS layer L_i in T and which is also the uppermost node of a fast segment F in T. If v receives a new message, then v is set to perform a transmission to its immediate fast child w in F in time step t', where $t' = i + 3j \bmod 3r_{max}$.

Note that since w is in L_{i+1} and has the same rank as v it follows that w will transmit this message to its fast child (if it is not the lowest node in F) in time step $t'' = t' + 1 \bmod 3r_{max}$. Thus it follows that after at most $3r_{max} + |F| - 1$ time steps the message will have reached every node in F.

(2) **Slow steps:** These are used to transmit messages across slow edges. Consider a node v on BFS layer L_i in $\tilde{T}$ that belongs to a slow segment of $\tilde{T}$. If v receives a new message, then v will perform a transmission only in time step t' satisfying $t' = i \bmod 3$. We employ the procedure RCW from [19] to perform the transmissions in the slow steps.

Procedure RCW allows the movement of one unit messages from one partition of a bipartite graph of size n' (here, an entire BFS layer L_i of $\tilde{T}$) to the other (here, the next layer L_{i+1}) with high probability in time $O(\log n')$. Note that as soon as the message has reached a (slow) node w in layer L_{i+1} then w will start to transmit this to its (slow) descendants (within 3 time units).

The probability that the procedure RCW is successful in transmitting a message between two nodes is given by the the following lemma from [19].

Lemma 5. *The probability that a node v in layer L_i will be successful in transmitting a message to its adjacent nodes in L_{i+1} is given by $p \geq 1/(4e)$.*

Note that the described pattern of transmissions separates the transmissions between the fast and slow steps by at least one unit of time. The pattern also ensures that at any time step, transmissions are performed on BFS layers at distances that are multiples of 3 apart. Thus there will be no conflicts between transmissions coming from different BFS layers. There is also no collisions in transmissions between the fast nodes of $\tilde{T}$ with the same rank and within the same BFS layer of T due to the properties of a gathering spanning tree (see [19]).

Corollary 1. *The total time spent on fast transmissions in sending a unit message from c to a leaf v is $O(\log^2 n + r)$ where r is the radius of G.*

Proof. As stated, the time to traverse a fast segment F takes time $3r_{max} + |F| - 1 = O(\log n + |F|)$. There are at most $\log n$ fast segments on the path from c to v in T and the sum of the lengths of these segments is at most r, thus the result follows.

Since the construction of a gathering spanning tree T (and consequently of $\tilde{G}$ and $\tilde{T}$) is polynomial in view of Lemma 2.2, we have the following result.

Theorem 2. *For any n-node graph G, there exists a randomized gossiping algorithm with unit size messages that runs in time polynomial in n.*

It remains to bound the probability of success of our gossiping schedule and to estimate the length of the scheme constructed by it.

Theorem 3. *There exists a randomized algorithm that for any known topology radio network of size n, following a polynomial time preprocessing stage, solves the gossiping problem with unit size messages with high probability in time $O(n \log n)$.*

Proof. Consider an arbitrary node v in the graph G, and consider the path along which it is supposed to get the message from the root. This path is divided into "fast segments" and "slow segments" as discussed above. Consider first the fast segments on the path from c to a leaf v and let F be the topmost segment on this path and w the topmost node of F. Then it follows that the messages will be transmitted from w in time steps at most $O(\log n)$ apart. Thus the last message will be transmitted after $O(n \log n)$ time. >From Corollary 1 it follows that this message will spend $O(n + \log^2 n)$ time on fast transmissions before it reaches every descendant including v. Thus the total time spent on fast transmissions is bounded by $O(n \log n + \log^2 n)$.

We now claim that the total number of time steps spent by the messages for the slow steps on its way to v is at most $O(n \log n)$ as well. For each slow transmission we will be activating the RCW procedure $O(1)$ times to ensure that it reaches every node with high probability. Thus the number of times the RCW procedure is activated for a particular message is bounded by the height of $\tilde{T}$ which is $O(\log n)$. It follows that the total time spent on slow transmissions for one particular message is bounded by $O(\log^2 n)$. Since the slow transmissions are performed at intervals that are $O(1)$ apart the last message will start to be transmitted after $O(n \log n)$ time and reach every node after spending $O(\log^2)$ time on slow transmissions. Thus the combined time spent on both fast and slow transmissions is bounded by $O(n \log n)$.

It remains to show that each message will reach every node with high probability. Note first that each participation of a particular message in an activation of the RCW procedure succeeds (i.e., the message crosses from its current node to the next node on the path to v) independently with constant probability $p \geq 1/(4e)$. Then let $\tilde{R}$ be the height of $\tilde{T}$ and X a random variable denoting the number of successes of a particular message on its way from c to v. I.e. X denotes the number of levels of $\tilde{T}$ that the message has crossed successfully. For each level that the message has to cross we will activate RCW a total of $24e$ times. Thus the maximum total number of times the message will participate in the RCW procedure is $24e\tilde{R}$, the expected value of X is $\mu = 6\tilde{R}$. Due to the way $\tilde{T}$ was constructed, we know that $\tilde{R}$ is bounded by $2\log n$.

Using the Chernoff bound, the probability $P_{fail}(v)$ that the message will not reach v after $24e\tilde{R}$ participations in RCW can be bounded from above by

$$P_{fail}(v) \leq P(X < \tilde{R}) = P(X < (1-5/6)\mu) < \exp\left(-\frac{1}{2}\left(\frac{5}{6}\right)^2 \mu\right) < n^{-1}.$$

Subsequently, the probability that the message will require more than $24e\tilde{R}$ participations in the RCW procedure before it reaches v, is smaller than $1/n$.

The presented algorithm requires the possibility to store n messages in the center node c. This happens just after finishing the gathering stage before the start of the broadcast stage. We note that it is possible to get around this requirement by interleaving the gathering and broadcasting stage. In this way no node would need more than $O(1)$ extra space. In fact, if y denotes the maximal number of simultaneous messages allowed in a receive or send buffer on any node then it is possible to modify the presented algorithm so that it solves the gossiping problem with high probability in time $O((\frac{n}{y} + r)\log n)$ where r is the radius of the network. Due to space limitation we defer more details to the full version of the paper.

3   Conclusion

We have proposed a new efficient (polynomial time) randomized schedule that performs the gossiping task with unit size messages in radio networks with high probability in optimal time $O(n \log n)$. The evident open problem regarding gossiping is whether there exists a deterministic gossiping schedule of time $O(n \log n)$ for every n-node graph G.

References

1. B. Awerbuch and D. Peleg. Sparse partitions. *Proc. 31st Symp. on Foundations of Computer Science*, 1990, pp. 503-513.
2. N. Alon, A. Bar-Noy, N. Linial and D. Peleg. A lower bound for radio broadcast. *J. Computer and System Sciences* 43, (1991), 290 - 298.
3. R. Bar-Yehuda, O. Goldreich and A. Itai. On the time complexity of broadcasting in radio networks: an exponential gap between determinism and randomization. *Proc. 5th Symp. on Principles of Distributed Computing*, 1986, 98 - 107.
4. M. Chrobak, L. Gąsieniec and W. Rytter, Fast broadcasting and gossiping in radio networks. *J. of Algorithms* 43(2), (2002), pp. 177-189.
5. I. Chlamtac and S. Kutten. On broadcasting in radio networks-problem analysis and protocol design. *IEEE Trans. on Communications* 33, (1985), pp. 1240-1246.
6. T.H. Cormen, C.E. Leiserson and R.L. Rivest. *Introduction to Algorithms.* MIT Press, 1990.
7. F. Cicalese F. Manne and Q. Xin. Faster centralized communication in radio networks. Manuscript, 2006.
8. A. Czumaj and W. Rytter. Broadcasting algorithms in radio networks with unknown topology. *Proc. 44th Symp. on Foundations of Computer Science*, 2003, pp. 492-501.
9. I. Chlamtac and O. Weinstein. The wave expansion approach to broadcasting in multihop radio networks. Proc. *Proc. INFOCOM*, 1987.
10. K. Diks, E. Kranakis and A. Pelc. The impact of knowledge on broadcasting time in radio networks. *Proc. 7th European Symp. on Algorithms*, 1999, pp. 41-52.
11. M. Elkin and G. Kortsarz. Polylogarithmic inapproximability of the radio broadcast problem. *Proc. APPROX*, 2004, LNCS 3122.

12. M. Elkin and G. Kortsarz. Improved broadcast schedule for radio networks. *Proc. 16th ACM-SIAM Symp. on Discrete Algorithms*, 2005.

13. I. Gaber and Y. Mansour. Broadcast in radio networks. *Proc. 6th ACM-SIAM Symp. on Discrete Algorithms*, 1995, pp. 577-585.

14. M. Christersson, L. Gąsieniec and A. Lingas. Gossiping with bounded size messages in ad-hoc radio networks. *Proc. 29th Int. Colloq. on Automata, Languages and Programming*, 2002, pp. 377-389.

15. L. Gąsieniec and A. Lingas. On adaptive deterministic gossiping in ad hoc radio networks, *Information Processing Letters* 2(83), 2002, pp. 89-94.

16. L. Gąsieniec, E. Kranakis, A. Pelc and Q. Xin. Deterministic M2M multicast in radio networks. *Proc. 31st Int. Colloq. on Automata, Languages and Programming*, 2004, LNCS 3142, pp. 670-682.

17. L. Gąsieniec and I. Potapov, Gossiping with unit messages in known radio networks. *Proc. 2nd IFIP Int. Conference on Theoretical Computer Science*, 2002, pp. 193-205.

18. L. Gąsieniec, I. Potapov and Q. Xin. Efficient gossiping in known radio networks. *Proc. 11th Int. Colloq. on Structural Information and Communication Complexity*, 2004, LNCS 3104, pp. 173-184.

19. L. Gąsieniec, D. Peleg and Q. Xin. Faster communication in known topology radio networks. *Proc. 24th Annual ACM SIGACT-SIGOPS Symposium on Principles of Distributed Computing*, PODC'2005, pp. 129-137.

20. L. Gąsieniec, T. Radzik and Q. Xin. Faster deterministic gossiping in ad-hoc radio networks. *Proc. 9th Scandinavian Workshop on Algorithm Theory*, 2004, LNCS 3111, pp. 397-407.

21. D. Kowalski and A. Pelc. Centralized deterministic broadcasting in undirected multi-hop radio networks. *Proc. APPROX*, 2004, LNCS 3122, pp. 171-182.

22. D. Liu and M. Prabhakaran. On randomized broadcasting and gossiping in radio networks. *Proc. 8th Int. Conf. on Computing and Combinatorics*, 2002, pp. 340-349.

23. A. Sen and M.L. Huson. A new model for scheduling packet radio networks. *Proc. 15th Joint Conf. of IEEE Computer and Communication Societies*, 1996, pp. 1116-1124.

24. P.J. Slater, E.J.Cockayne and S.T. Hedetniemi. Information dissemination in trees. *SIAM J. on Computing* 10, (1981), pp. 892–701.

25. A.N. Strahler. Hypsometric (area-altitude) analysis of erosional topology. *Bull. Geol. Soc. Amer.* 63, (1952), pp. 117–1142.

26. X.G. Viennot. A Strahler bijection between Dyck paths and planar trees. *Discrete Mathematics* 246, (2002), pp. 317–329.

27. Y. Xu. An $O(n^{1.5})$ deterministic gossiping algorithm for radio networks. *Algorithmica*, 36(1), (2003), pp. 93–96.

Author Index

Lecture Notes in Computer Science

For information about Vols. 1–4235

please contact your bookseller or Springer

Vol. 4275: R. Meersman, Z. Tari (Eds.), On the Move to Meaningful Internet Systems 2006: CoopIS, DOA, GADA, and ODBASE, Part I. XXXI, 1115 pages. 2006.

Vol. 4274: Q. Huo, B. Ma, E.-S. Chng, H. Li (Eds.), Chinese Spoken Language Processing. XXIV, 805 pages. 2006. (Sublibrary LNAI).

Vol. 4273: I. Cruz, S. Decker, D. Allemang, C. Preist, D. Schwabe, P. Mika, M. Uschold, L. Aroyo (Eds.), The Semantic Web - ISWC 2006. XXIV, 1001 pages. 2006.

Vol. 4272: P. Havinga, M. Lijding, N. Meratnia, M. Wegdam (Eds.), Smart Sensing and Context. XI, 267 pages. 2006.

Vol. 4271: F.V. Fomin (Ed.), Graph-Theoretic Concepts in Computer Science. XIII, 358 pages. 2006.

Vol. 4270: H. Zha, Z. Pan, H. Thwaites, A.C. Addison, M. Forte (Eds.), Interactive Technologies and Sociotechnical Systems. XVI, 547 pages. 2006.

Vol. 4269: R. State, S. van der Meer, D. O'Sullivan, T. Pfeifer (Eds.), Large Scale Management of Distributed Systems. XIII, 282 pages. 2006.

Vol. 4268: G. Parr, D. Malone, M. Ó Foghlú (Eds.), Autonomic Principles of IP Operations and Management. XIII, 237 pages. 2006.

Vol. 4267: A. Helmy, B. Jennings, L. Murphy, T. Pfeifer (Eds.), Autonomic Management of Mobile Multimedia Services. XIII, 257 pages. 2006.

Vol. 4266: H. Yoshiura, K. Sakurai, K. Rannenberg, Y. Murayama, S. Kawamura (Eds.), Advances in Information and Computer Security. XIII, 438 pages. 2006.

Vol. 4265: L. Todorovski, N. Lavrač, K.P. Jantke (Eds.), Discovery Science. XIV, 384 pages. 2006. (Sublibrary LNAI).

Vol. 4264: J.L. Balcázar, P.M. Long, F. Stephan (Eds.), Algorithmic Learning Theory. XIII, 393 pages. 2006. (Sublibrary LNAI).

Vol. 4263: A. Levi, E. Savaş, H. Yenigün, S. Balcısoy, Y. Saygın (Eds.), Computer and Information Sciences – ISCIS 2006. XXIII, 1084 pages. 2006.

Vol. 4262: K. Havelund, M. Núñez, B. Wolff, G. Roşu (Eds.), Formal Approaches to Software Testing and Runtime Verification. VIII, 255 pages. 2006.

Vol. 4261: Y. Zhuang, S. Yang, Y. Rui, Q. He (Eds.), Advances in Multimedia Information Processing - PCM 2006. XXII, 1040 pages. 2006.

Vol. 4260: Z. Liu, J. He (Eds.), Formal Methods and Software Engineering. XII, 778 pages. 2006.

Vol. 4259: S. Greco, Y. Hata, S. Hirano, M. Inuiguchi, S. Miyamoto, H.S. Nguyen, R. Słowiński (Eds.), Rough Sets and Current Trends in Computing. XXII, 951 pages. 2006. (Sublibrary LNAI).

Vol. 4257: I. Richardson, P. Runeson, R. Messnarz (Eds.), Software Process Improvement. XI, 219 pages. 2006.

Vol. 4256: L. Feng, G. Wang, C. Zeng, R. Huang (Eds.), Web Information Systems – WISE 2006 Workshops. XIV, 320 pages. 2006.

Vol. 4255: K. Aberer, Z. Peng, E.A. Rundensteiner, Y. Zhang, X. Li (Eds.), Web Information Systems – WISE 2006. XIV, 563 pages. 2006.

Vol. 4254: T. Grust, H. Höpfner, A. Illarramendi, S. Jablonski, M. Mesiti, S. Müller, P.-L. Patranjan, K.-U. Sattler, M. Spiliopoulou, J. Wijsen (Eds.), Current Trends in Database Technology – EDBT 2006. XXXI, 932 pages. 2006.

Vol. 4253: B. Gabrys, R.J. Howlett, L.C. Jain (Eds.), Knowledge-Based Intelligent Information and Engineering Systems, Part III. XXXII, 1301 pages. 2006. (Sublibrary LNAI).

Vol. 4252: B. Gabrys, R.J. Howlett, L.C. Jain (Eds.), Knowledge-Based Intelligent Information and Engineering Systems, Part II. XXXIII, 1335 pages. 2006. (Sublibrary LNAI).

Vol. 4251: B. Gabrys, R.J. Howlett, L.C. Jain (Eds.), Knowledge-Based Intelligent Information and Engineering Systems, Part I. LXVI, 1297 pages. 2006. (Sublibrary LNAI).

Vol. 4250: H.J. van den Herik, S.-C. Hsu, T.-s. Hsu, H.H.L.M. Donkers (Eds.), Advances in Computer Games. XIV, 273 pages. 2006.

Vol. 4249: L. Goubin, M. Matsui (Eds.), Cryptographic Hardware and Embedded Systems - CHES 2006. XII, 462 pages. 2006.

Vol. 4248: S. Staab, V. Svátek (Eds.), Managing Knowledge in a World of Networks. XIV, 400 pages. 2006. (Sublibrary LNAI).

Vol. 4247: T.-D. Wang, X. Li, S.-H. Chen, X. Wang, H. Abbass, H. Iba, G. Chen, X. Yao (Eds.), Simulated Evolution and Learning. XXI, 940 pages. 2006.

Vol. 4246: M. Hermann, A. Voronkov (Eds.), Logic for Programming, Artificial Intelligence, and Reasoning. XIII, 588 pages. 2006. (Sublibrary LNAI).

Vol. 4245: A. Kuba, L.G. Nyúl, K. Palágyi (Eds.), Discrete Geometry for Computer Imagery. XIII, 688 pages. 2006.

Vol. 4244: S. Spaccapietra (Ed.), Journal on Data Semantics VII. XI, 267 pages. 2006.

Vol. 4243: T. Yakhno, E.J. Neuhold (Eds.), Advances in Information Systems. XIII, 420 pages. 2006.

Vol. 4242: A. Rashid, M. Aksit (Eds.), Transactions on Aspect-Oriented Software Development II. IX, 289 pages. 2006.

Vol. 4241: R.R. Beichel, M. Sonka (Eds.), Computer Vision Approaches to Medical Image Analysis. XI, 262 pages. 2006.

Vol. 4239: H.Y. Youn, M. Kim, H. Morikawa (Eds.), Ubiquitous Computing Systems. XVI, 548 pages. 2006.

Vol. 4238: Y.-T. Kim, M. Takano (Eds.), Management of Convergence Networks and Services. XVIII, 605 pages. 2006.

Vol. 4237: H. Leitold, E. Markatos (Eds.), Communications and Multimedia Security. XII, 253 pages. 2006.

Vol. 4236: L. Breveglieri, I. Koren, D. Naccache, J.-P. Seifert (Eds.), Fault Diagnosis and Tolerance in Cryptography. XIII, 253 pages. 2006.